AF553596

Forest Genetics and Tree Breeding

Forest Genetics and Tree Breeding

Dr. Pradeep Kumar Choda

RANDOM PUBLICATIONS
NEW DELHI (INDIA)

Forest Genetics and Tree Breeding

ISBN 978-93-5111-830-5

Published in 2016 in India by

RANDOM PUBLICATIONS

4376-A/4B, Gali Murari Lal, Ansari Road
New Delhi-110 002
Phone : +9111-43580356, 011-23289044, 011-43142548
e-mail: sales@randompublications.com,
info@randompublications.com, randomexports@gmail.com
Reprint 2022

Type Setting by : Friends Media, Delhi-110089
Digitally Printed at : Replika Press Pvt. Ltd.

Preface

Forest genetic resources or tree genetic resources are genetic material of shrub and tree species of actual or future value. Forest genetic resources are essential for forest-depending communities who rely for a substantial part of their livelihoods on timber and non-timber forest products (for example fruits, gums and resins) for food security, domestic use and income generation. These resources are also the basis for large-scale wood production in planted forests to satisfy the worldwide need for timber and paper. Genetic resources of several important timber, fruit and other non-timber tree species are conserved ex situ in genebanks or maintained in field collections. Nevertheless in situ conservation in forests and on farms is in the case of most tree species the most important measure to protect their genetic resources.

Today, managing and conserving forest genetic resources are essential components of forest stewardship, contributing to B.C.'s position in a globally competitive forest industry. The Ministry of Forests and Range Forest Genetics research program includes not only tree improvement, but a wide range of initiatives related to genetic conservation, genetic resources management, and identifying and developing mitigation strategies for climate change impacts. A typical forest tree breeding program starts with selection of superior phenotypes (plus trees) in a natural or planted forest. This application of mass selection improves the mean performance of the forest. Offspring is obtained from selected trees and grown in test plantations that act as genetic trials. Based on such tests the best genotypes among the parents can be selected. Selected trees are typically multiplied by either seeds or grafting and seed orchards are established when the preferred output is improved seed. The topics of this book have so chosen as to give a comprehensive idea about forest genetics and tree breeding. The present book has been designed as a text-cum-reference book for post-graduate students, research workers, teachers and undergraduate students specializing in advance course of forest genetics and tree improvement.

– Author

Contents

1

Introduction

GMOs are defined as organisms that have been modified by the application of recombinant DNA technology (where DNA from one organism is transferred to another organism). The term "transgenic trees" is also used for GM trees, where a foreign gene (a transgene) is incorporated into the tree genome. One of the first reported trials with GM forest trees was initiated in Belgium in 1988 using poplars. A study carried out in 1999 indicated that, since then, there have been more than 100 reported trials, involving at least 24 tree species-most of which are timber-producing species.

The majority of the field trials were carried out in the USA and Canada. Whereas it is estimated that roughly 40 million hectares of transgenic agricultural crops were grown commercially in 1999 (ISAAA figures), there is no reported commercial-scale production of transgenic trees. Information on field trials of GM trees has been published by the OECD and the World Wide Fund for Nature (1999). Traits for which genetic modification can realistically be contemplated in the near future include insect and virus resistance, herbicide tolerance and lignin content.

However, insertion of any gene into a tree species with expected functional results will be a substantial undertaking and insertion of enough genes to confer *e.g.* long-term insect resistance in a perennial species even more so. Virus and insect resistance, in particular, are of major significance for crop plants. By contrast, these traits are not the most important in breeding programmes of forest tree species (poplars being an exception). Reduction of lignin is a valuable objective for species producing pulp for the paper industry; work on this aspect is underway in aspen.

A major technical factor limiting the application of genetic modification to forest trees, is the current low level of knowledge regarding the molecular control of traits which are of most interest, notably those relating to growth and stem and wood quality. Genetic modification of these traits remains a distant prospect. Investments in genetic modification technologies should be weighed against the possibilities of exploiting the large amounts of genetic variation, which are generally untapped, available within any single species in nature.

Biosafety aspects of GM trees need careful consideration because of the long generation time of trees, their important role in ecosystem functioning and the potential for long distance dispersal of pollen and seed.

PRINCIPLES OF GENETICS

Mendel's studies have provided scientists with the basis for mathematically predicting the probabilities of genotypes and phenotypes in the offspring of a genetic cross. But not all genetic observations can be explained and predicted based on Mendelian genetics. Other complex and distinct genetic phenomena may also occur. Several complex genetic concepts, described in this section, explain such distinct genetic phenomena as blood types and skin colour.

INCOMPLETE DOMINANCE

In some allele combinations, dominance does not exist. Instead, the two characteristics blend. In such a situation, both alleles have the opportunity to express themselves. For instance, snapdragon flowers display incomplete dominance in their colour. There are two alleles for flower colour: one for white and one for red. When two alleles for white are present, the plant displays white flowers. When two alleles for red are present, the plant has red flowers. But when one allele for red is present with one allele for white, the colour of the snapdragons is pink.

However, if two pink snapdragons are crossed, the phenotype ratio of the offspring is one red, two pink, and one white. These results show that the genes themselves remain independent; only the expressions of the genes blend. If the gene for red and the gene for white actually blended, pure red and pure white snapdragons could not appear in the offspring.

Multiple Alleles

In certain cases, more than two alleles exist for a particular characteristic. Even though an individual has only two alleles, additional alleles may be present in the population. This condition is multiple alleles. An example of multiple alleles occurs in blood type. In humans, blood groups are determined by a single gene with three possible alleles: A, B, or O. Red blood cells can contain two antigens, A and B. The presence or absence of these antigens results in four blood types: A, B, AB, and O. If a person's red blood cells have antigen A, the blood type is A. If a person's red blood cells have antigen B, the blood type is B. If the red blood cells have both antigen A and antigen B, the blood type is AB. If the red blood cells have neither antigen A nor antigen B, the blood type is O. The alleles for type A and type B blood are co-dominant; that is, both alleles are expressed. However, the allele for type O blood is recessive to both type A and type B. Because a person has only two of the three alleles, the blood type varies depending on which two alleles are present. For instance, if a person has the A allele and the B allele, the blood type is AB. If a person has two A alleles, or one A and one O allele, the blood type is A. If a person has two B alleles or one B and one O allele, the blood type is B. If a person has two O alleles, the blood type is O.

Polygenic Inheritance

Although many characteristics are determined by alleles at a single place on the chromosome, some characteristics are determined by an interaction of genes on several chromosomes or at several places on one chromosome. This condition is polygenic inheritance. An example of polygenic inheritance is human skin colour. Genes for skin colour are located in many places, and skin colour is determined by which genes are present at these multiple locations. A person with many genes for dark skin will have very dark skin colour, and a person with multiple genes for light skin will have very light skin colour. Many people have some genes for light skin and some for dark skin, which explains why so many variations of skin colour exist. Height is another characteristic probably reflecting polygenic characteristics.

Gene linkage

A chromosome has many thousands of genes; there are an estimated 100,000 genes in the human genome. Inheritance involves the transfer of chromosomes from parent to offspring through meiosis and sexual reproduction. It is common for a large number of genes to be inherited together if they are located on the same chromosome. Genes that are inherited together are said to form a linkage group. The concept of transfer of a linkage group is gene linkage. Gene linkage can show how close two or more genes are to one another on a chromosome. The closer the genes are to each other, the higher the probability that they will be inherited together. Crossing over occurs during meiosis, but genes that are close to each other tend to remain together during crossing over.

Sex Linkage

Among the 23 pairs of chromosomes in human cells, one pair is the sex chromosomes. (The remaining 22 pairs of chromosomes are referred to as autosomes.) The sex chromosomes determine the sex of humans. There are two types of sex chromosomes: the X chromosome and the Y chromosome. Females have two X chromosomes; males have one X and one Y chromosome. Typically, the female chromosome pattern is designated XX, while the male chromosome pattern is XY. Thus, the genotype of the human male would be 44 XY, while the genotype of the human female would be 44 XX (where 44 represents the autosomes). In humans, the Y chromosome is much shorter than the X chromosome. Because of this shortened size, a number of sex-linked conditions occur. When a gene occurs on an X chromosome, the other gene of the pair probably occurs on the other X chromosome. Therefore, a female usually has two genes for a characteristic. In contrast, when a gene occurs on an X chromosome in a male, there is usually no other gene present on the short Y chromosome. Therefore, in the male, whatever gene is present on the X chromosome will be expressed.

DNA Defined

During the 1950s, a tremendous explosion of biological research occurred, and the methods of gene expression were elucidated. The knowledge generated during this period helped explain how genes function, and it gave rise to the science of molecular genetics. This science is based on the activity of deoxyribonucleic acid (DNA) and how this activity brings about the production of proteins in the cell. Genetic material is packaged into DNA molecules. DNA molecules relay the inherited information to messenger RNA (mRNA) which, in turn, codes for proteins. This chain of command is represented as:

$$\text{DNA} \rightarrow \text{mRNA} \rightarrow \text{Protein}$$

The flow of information from DNA to protein is known as the *Central Dogma* of molecular biology. In 1953, two biochemists, James D. Watson and Francis H.C. Crick, proposed a model for the structure of DNA. (In 1962, they shared a Nobel Prize for their work.) The publication of the structure of DNA opened a new realm of molecular genetics. Its structure provided valuable insight into how genes operate and how DNA can reproduce itself during mitosis, thereby passing on hereditary characteristics. Not only did the new research uncover many of the principles of protein synthesis, but it also gave rise to the science of biotechnology and genetic engineering.

DNA Structure

As proposed by Watson and Crick, deoxyribonucleic acid (DNA) consists of two long nucleotide chains. The two nucleotide chains twist around one another to form a double helix, a shape resembling a spiral staircase. Weak chemical bonds between the chains hold the two chains of nucleotides to one another.

A nucleotide in the DNA chain consists of three parts: a nitrogenous base, a phosphate group, and a molecule of deoxyribose. The nitrogenous bases of each nucleotide chain are of two major types: purines and pyrimidines. Purine bases have two fused rings of carbon and nitrogen atoms, while pyrimidines have only one ring. The two purine bases in DNA are adenine (A) and guanine (G). The pyrimidine bases in DNA are cytosine (C) and thymine (T). Purines and pyrimidine bases are found in both strands of the double helix. The phosphate group of DNA is derived from a molecule of phosphoric acid. The phosphate group connects the deoxyribose molecules to one another in the nucleotide chain. Deoxyribose is a five-carbon carbohydrate. The purine and pyrimidine bases are attached to the deoxyribose molecules, and the purine and pyrimidine bases are opposite one another on the two nucleotide chains. Adenine is always opposite thymine and binds to thymine. Guanine is always opposite cytosine and binds to cytosine. Adenine and thymine are said to be complementary, as are guanine and cytosine This is known as the principle of complementary base pairing.

DNA Replication

Before a cell enters the process of mitosis, its DNA replicates itself. Equal copies of the DNA pass into the daughter cells at the end of mitosis. In human cells, this means that 46 chromosomes (or molecules of DNA) replicate to form 92 chromosomes. The process of DNA replication begins when specialised enzymes pull apart, or "unzip," the DNA double helix. As the two strands separate, the purine and pyrimidine bases on each strand are exposed. The exposed bases then attract their complementary bases. Deoxyribose molecules and phosphate groups are present in the nucleus. The enzyme DNA polymerase joins all the nucleotide components to one another, forming a long strand of nucleotides. Thus, the old strand of DNA directs the synthesis of a new strand of DNA through complementary base pairing. The old strand then unites with the new strand to reform a double helix. This process is called *semiconservative replication* because one of the old strands is conserved in the new DNA double helix.

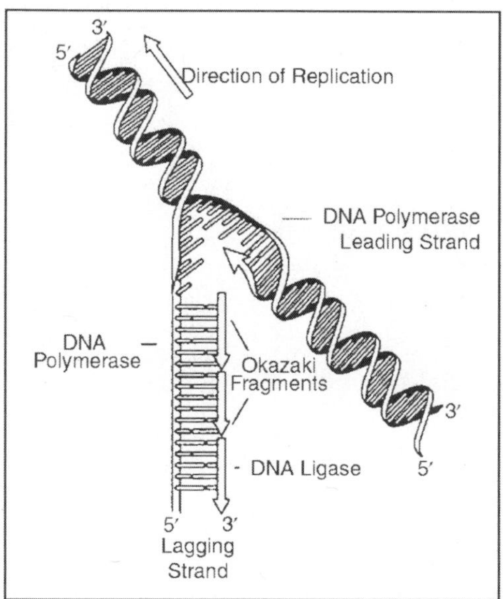

Fig. DNA replication. The double helix opens and a complementary strand of DNA is synthesised along each strand.

DNA polymerase joins nucleotides in a 52-32 direction on the leading strand, shown in Figure . However, DNA polymerase does not elongate a DNA strand in a 32-52 direction. Therefore, the 32-52 strand, called the lagging strand, is synthesised in short segments in a 52-32 direction. These short segments placed on the lagging strand are Okazaki fragments and are ultimately joined together by the enzyme DNA ligase to form a new DNA strand. DNA replication occurs during the S phase of the cell cycle. After replication has taken place, the chromosomal material shortens and thickens. The chromatids appear in the prophase of the next mitosis. The process then continues, and eventually two daughter cells form, each with the identical amount and kind of DNA as the parent cell. The process of DNA replication thus ensures that the molecular material passes to the offspring cells in equal amounts and types

PROTEIN SYNTHESIS

During the 1950s and 1960s, it became apparent that DNA is essential in the synthesis of proteins. Proteins are used in enzymes and as structural materials in cells. Many specialised proteins function in cellular activities. For example, in humans, the hormone insulin and the muscle cell filaments are composed of protein. The hair, skin, and nails of humans are composed of proteins, as are all the hundreds of thousands of enzymes in the body The key to a protein molecule is how the amino acids are linked. The sequence of amino acids in a protein is a type of code that specifies the protein and distinguishes one protein from another. A genetic code in the DNA determines this amino acid code. The genetic code consists of the sequence of nitrogenous bases in the DNA. How the nitrogenous base code is translated to an amino acid sequence in a protein is the basis for protein synthesis.

For protein synthesis to occur, several essential materials must be present, such as a supply of the 20 amino acids, which comprise most proteins. Another essential element is a series of enzymes that will function in the process. DNA and another form of nucleic acid called ribonucleic acid (RNA) are essential. RNA is the nucleic acid that carries instructions from the nuclear DNA into the cytoplasm, where protein is synthesised. RNA is similar to DNA, with two exceptions. First, the carbohydrate in RNA is ribose rather than deoxyribose, and second, RNA nucleotides contain the pyrimidine uracil rather than thymine.

FOREST TREES

The earth has been gutted by thought-less rape of the soil, afforestation is one of the best and quickest ways to stop erosion and to build up another layer of water- retaining and wind-defying topsoil. This is true almost every- where, even in places where it would seem that the restoration of something like the original grass-mat would be a better solution. The reason for this lies in the circumstance that grass fires and over-grazing may spoil the prairie more easily than the forest. In certain cases trees are a valuable crop, counting in production per unit of area. In some countries the same land may produce more profit if we cut down its eucalyptus for railway sleepers every ten years than if we had ten agricultural crops. In countries with poorish soil, as in parts of Canada and Sweden, spruce and poplar for paper-pulp give a more paying crop than the cereal crops that could be grown there.

Special forestry breeding stations have been established, notably in the U.S.A. and in Sweden, to study the quality of certain trees and methods of propagation and plant breeding. There are many countries where a study of forest trees and of breeding trees has not been made the lifework of foresters and geneticists in special experiment stations. There is one curious fact which strikes every biologist. In general, we find that if a species has a considerable area, so that it is common in many different countries and latitudes, the trees that thrive best and give us the most profitable forests are the local trees. The best seed is generally that which is collected in our own forests. This is very striking in a comparison of Finnish, Dutch and French pines, when tried out in Finnish, Dutch and French plantings. In every case the trees from home-grown seeds are the most valuable ones in the country where several "provenances" are tested.

In the wild state trees are almost without exception propagated from seeds, but in nurseries vegetative reproduction is often used. This makes it possible to reproduce valuable trees, even if they happen to be heterozygous, and often the trees produced by layering are as large and strong as seedlings of several years of age.

Budding on seedlings costs too much in forestry work, but it is now possible to make cuttings root in species from which such cuttings could not be induced to grow ten years ago. Treatment of the cuttings with chemicals sometimes helps. Occasionally mutations are found in wild trees, and a good example of a very valuable mutation in trees is that of the triploid aspen discovered by Nilsson-Ehle in Sweden. Some triploids make phenomenal growth, and the same is true of some species hybrids in the groups of the pines and of the eucalypts. In the absence of a cheap method of vegetative reproduction the utilization of hybrids becomes almost impossible.

When forest trees must be sown, the first requirement is to find a good source of seed. In this respect some species of forest trees have the advantage that the older, selected trees in old forests are very good seed producers. In other species the cheapest seed is produced from young, widely spaced, poorly shaped trees, and it is almost impossible to judge such trees even as individuals. A system of progeny testing in the female line, with phaenotypic selection of parent trees, is about the most efficient system of selection in such trees as beech and oak. This is a very long-drawn- out process. After all, plantings of trees along the roads may often give us stands that are practically isolated from free intercrossing, and it would be quite possible to obtain very valuable data for future work if some system of book-keeping would enable us to remember that a certain set of trees was derived from certain specified trees. In very early flowering and very variable trees, like Robinia, some Eucalypts, Hevea and birches, it would be quite feasible to inaugurate a system of selection according to progeny tests that would work just as well as that worked out for sugar-beets. In afforestation it is sometimes advantageous to use non-indigenous genera.

Eucalypts and wattles have spread all over the world in appropriate climates. Great care should be taken, in the importation of nursery stock, to exclude the possibility of introducing pests, micro-organisms or virus diseases that would affect those trees and might spread to other, native stands. On the other hand, it is sometimes possible to find a parasite that helps us in the fight against a pest, if we look for this in the country of origin (cactoblastis), The circumstance that it takes so long to grow a forest to maturity brings its very special difficulties and dangers.

Attempts to judge immature trees, or even seedlings, by qualities that we think correlates them with the really valuable essential characters which we appreciate in mature trees. It is very tempting to do so: to look for upright shape in oak seedlings and for very quick growth in poplar or spruce grown for pulpwood. Undoubtedly some of those valuable qualities ought to be appreciable in quite young stands, but correlation should be proved to exist, and not just taken for granted. After all, the balance of good and bad qualities determines the final value of a forest, and the environment, soil, climate and parasites must be taken into account more than in any other and shorter-lived crop.

If we use the wrong kind of seed, because it is cheap, or because it looks plump and clean, or because it germinates well, our contemporaries may applaud us, while our grand- children will curse us for our folly. One rule should never be lost sight of: no seed should ever be used for afforestation work unless we have a guarantee that the trees on which it is produced are of very good quality, grown in conditions for which the seedlings are intended.

This guarantee can never be given for imported seed, and perhaps it is a safe rule never to use imported seed for any other than experimental purposes. Of course it may be pointed out that imported seed, even from exotic species, has given us extremely valuable additions to local floras. But even in such cases as that of Eucalyptus, of Oregon pines, of the oil-palm, large-scale plantings without preliminary experimentation have been undertaken by planters or by companies that had the courage to take risks. Such risks have often enough resulted in costly failures. It is an excellent plan to profit by the successes of the pioneers, and in the absence of well-equipped and well-endowed experiment stations this is the logical way in which all human culture must progress. We can use a neighbouring country's pioneering work in practice (or in science for that matter) as a cheap substitute for local effort, especially in the growing of such long-lived crops as trees.

In so far as the narrower field of the efforts of the geneticist- plant-breeders is concerned, the "hybrid vigour" so strikingly shown in some tree hybrids can be taken full advantage of only where we can either multiply our stocks by some method of vegetative reproduction that is cheap enough for planting forests, or where we can produce first-generation hybrids by routine methods. The investigation of the methods of propagation is such a very important point that this alone would merit the founding of forestry experiment stations. If we compare the value of a mature tree with the cost of the dozen seeds that were used to grow the treelets from which this tree survived, we see that not very much can be gained by using cheap seed or cheap nursery stock.

Citrus

Plant-breeding in the citrus group has so many curious aspects. To a great extent some of the very peculiar problems were solved when it was discovered that every citrus seed may contain three to ten embryos, and that of those embryos only one is derived from a fertilized ovum. All those supernumerary plantlets that will grow out of the seed are derived from the maternal tissue, so that sowing the seed will often give us vegetative reproduction.

Swingle and others discovered that there is a discrepancy between the variability in lots of seedlings from self-fertilized seed and that in lots of seedlings after crossing. Self fertilization in most other plants that are as heterozygous as our common oranges and lemons gives us exceedingly variable offspring. But here it does not! The probable solution lies in the great "hybrid vigour" of cross-bred plants, that win the competition with the purely maternal sprouts, while the self-fertilized "real" seedlings lose the same competition.

The importation of foreign clones can easily be accomplished by means of seeds, and this obviates taking any risks of importing pests or virus diseases. It necessitates separating the plantlets that grow from one seed. Even seedless oranges, grapefruit or mandarines can be propagated from seed, there is any exception to the rule that even well- known seedless varieties will occasionally give us a viable seed.

Not only are the widest, wildest crosses possible in Citrus, but the proportion of interesting novelties among hybrids is very great. For this reason, citrus is excellent material for the amateur. Some experimental citrus gardens are really delightful places! On the other hand, the exigencies of the fruit trade are such that one or at the utmost two varieties will suffice in any orange or grapefruit-exporting country. To give the public what it knows, is more important than to give it something of superior quality. Lemons can be crossed with oranges, and either of them with grape-fruit. Many hybrids between mandarines and grapefruit happen to be among the very best oranges! And so it goes. It would not be too difficult to produce oranges of sixteen to the two-cubic-feet box, but some markets prefer quite small ones to large ones.

Curiously enough, some of the best varieties may fit conditions in very diverse circumstances, witness the history of the Brazilian Washington Navel orange. In citriculture much can be accomplished by careful importation and comparison of many commercial exotic varieties in collections. There is no good reason for trying to establish an export trade in some of the fifth-rate oranges and insipid tangerines that some countries are producing to-day. Even should the export of nursery stock or buds be prohibited or frowned upon, seeds are always exported in the fruit in enormous quantities, and such seeds will certainly reproduce the variety, even if they sometimes give us some novel material in addition to the old.

Disease-resistant and frost-resistant species can be combined in hybrids with commercial citrus groups, and very valuable results should be obtained unless the plant breeder lacks the patience or the insight to value his first generation hybrids even if they fail to be of real individual value. An excellent orange or lemon, adapted to a new citrus-growing country in the tropics, the importation of a few excellent subtropical widely known varieties. Much prefer to collect nursery stock, or preferably fruit in old citrus- growing districts in the tropics, where this could be obtained in very mixed (and perhaps on the whole inferior) orchards.

FORESTRY IN DEVELOPING COUNTRIES

Forests cover approximately 30 percent of the world's total land area. They are the source of vital commodities, including raw materials and food and are essential for maintaining agricultural productivity and the environmental well-being of the planet as a whole. They protect soil and water and buffer the effects of wind and rain, thus helping to decrease soil erosion and they are an important sink for carbon dioxide. Forests are also among the most important repositories of biological diversity. Roughly 500 million rural people live in, or close to, forests. Most communities use a variety of forest products, particularly those in developing countries. Plant stems, tubers and fruits provide additional food during hungry seasons or when crops fail; wild animals are harvested for meat and hides; and the forests provide fuelwood, fodder for livestock, medicines and other products and services. The most important trend in forestry in developing countries is the progressive reduction in the area of forests due to changes in land use. Another important trend, evident at a global level, is increasing forest degradation through unmanaged use. When forests are degraded, their productive functions and their capacity as regulators of the environment are reduced, increasing flood and erosion hazards, reducing soil fertility and contributing to the loss of forest products and overall loss of biological diversity.

While forests are being lost, there is growing demand both for environmental services and for wood and wood products which they provide. A forecast by FAO predicts that wood demand is expected to increase by 25 percent from 1996 to 2010. This demand will, increasingly, have to be met by forest plantations, and with decreasing land areas available for forestry, plantation methods will have to be increasingly intensive. This will necessitate better tree improvement programmes in which biotechnology may play a role.

THE WEALTH OF INTRAPOPULATION GENETIC VARIATION

Within populations, four forces (selection, migration, mutation and population size) determine the structure of genetic variation. The interaction of all four factors can be highly complex and their effects are determined by the mating system and the kinds of gene effects that exist. A problem in analysing even this level of genetic evolution is that the kinds of gene effects that exist depend on the effects that other forces of evolution have had on gene actions themselves. To consider how the forces can affect evolution, we can first start with each of them and then consider how their interactions can change predictions.

For selection to have an effect on speciation and the structure of forests, there must be a difference among genotypes in the traits affected by survival or reproduction events and the individuals must accordingly leave more or fewer progeny than other genotypes. The genotype is defined as the individuals with a distinctive ensemble of alleles at its gene loci that distinguishes it from other genotypes. This is a multiple gene locus definition, which is often simplified by considering every one of the loci to be independent in the occurrence of its alleles in both frequency and effect.

In this case, we can consider the genotype to be the summation of all independent loci and hence can examine gene effects and dynamics one locus at a time. This is, of course, an oversimplified view since genes are in fact linked on chromosomes and therefore have linkages that correlate their occurrences, and most genes affect processes that are combined in their net effect on traits; furthermore, most genes affect enzymatic pathways that ultimately affect several traits when measured at the level of whole organism performance. These effects are called linkage, epistasis and pleiotropy, respectively, and tie the whole organism together into an integrated set of actions and functions that can survive and reproduce. While undoubtedly true in general, there is little evidence that individuals are so bound by such constrained genetic constitutions and physiological limitations that substantial independence among loci exists. The evidence for the existence of strongly selected, tightly linked 'coadapted gene complexes' is not strong and hence the model of independence can be usefully employed, even though it is surely wrong at least in detail.

Within a gene locus, many alleles may exist and the pairwise interactions among them are called the dominance relations. For a diploid organism, if both alleles at a locus are the same, the individual is said to be a homozygote; if the two alleles are different, the individual is said to be a heterozygote. If the heterozygote has a phenotype intermediate between the two homozygotes, the dominance effect is considered to be zero and the gene action is said to be additive. If the heterozygote is other than intermediate, it may be less than the lesser of the two homozygotes. If its phenotype lies between the homozygotes but is not exactly intermediate and if it appears to be greater than the greater of the homozygotes, the gene action is said to be underdominant, partially dominant or overdominant, respectively.

The effect of selection would depend on the kind of gene action, since directional selection for greater or lesser phenotypes would favour different homozygote and heterozygote states. Selection may also be stabilizing if it favours an intermediate phenotype or disruptive if it disfavours the intermediate and its effect on the frequency of the alleles would depend on the gene actions. Directional selection experiments seem to indicate that most alleles have partial dominance actions, though many may be involved in multiple locus interactions. Since selection would be expected to favour survival and reproduction of certain genotypes, we expect that it would increase the population level of fitness. Selection is therefore usually considered to have a positive effect on the evolutionary potential of a population or species in small increments every generation but may not generate an optimal condition at any one time. For alleles that have partial dominance effects, the dynamics of the rates of displacement of one allele by another are such that the frequency of the allele itself affects its own rate of displacement. When alleles are at either low or high frequency, their rates of change are slower than when they are at an intermediate frequency. Except in small populations where accidental loss can occur, selection would not generally lose alleles that are favoured for adaptation.

Another force of evolution is mutation, which is usually considered to be a random change in gene structure and function that occurs indiscriminately throughout the genome. Since most species have evolved a genetic and physiological level that is close to a sufficient capacity for continued evolution, novel changes are not expected to improve performance or fitness but rather to generally decrease adaptability. However, there are always going to be mutations and while most may be disfavoured by selection, there is likely to be a balance between the rate at which new mutants are introduced and the rate of their elimination. Most mutations occur at low rates for any one locus in any one generation and would not be expected to affect the adaptability of a population unless they are accumulated in small populations.

For most conifer species where the frequency of deleterious mutants has been estimated, there are a larger number of lethal alleles carried at one or another locus in most individual trees than in most other plant or animal species. This might indicate that there are many effective loci in trees that can mutate to a harmful state, that the mutation rate is high or that the mutants are favoured in some conditions or are of some advantage in heterozygous conditions. However, independent mutations and selection can generate differences among populations, which can increase fitness in each. Hybridization among populations can then generate high levels of genetic variance in traits that may then be useful in new environments.

A third force of evolution is migration, which can occur among many different kinds of population structures. The input of pollen or seed from other populations that have different allele frequencies may bring in alleles that might otherwise be lost, and if the migration rate is high enough among all populations may induce all populations to have approximately the same frequencies. If the immigrant alleles are favoured, then further immigration would advance the rate of its increase; if disfavoured, immigration would maintain alleles at higher frequencies than expected in a manner similar to mutation; and if neither, would tend to make frequencies homogeneous among all populations. In the presence of divergent selection among populations that exist in different environments, selection would act to eliminate migrant alleles and would maintain population divergence. The fourth force is sampling error induced by small population sizes, which tends to increase the probability that low-frequency alleles are lost by chance. If populations remain at low sizes for several generations, the chances of random loss are accumulated; even populations of more than 20-50 adults can lose alleles that may have started with moderate frequencies. Since most forest trees seem to carry many deleterious mutants, even if at low frequency at any one locus, another problem is that small populations will suffer from inbreeding depression and may eventually suffer such debilitation from several mildly depressive mutants that the population goes extinct. Individual populations would have to be large enough to sustain viability, and to avoid environmental accidents and catastrophes if no migrants were recruited.

There are two problems faced in persistently small populations, the loss of alleles and inbreeding depression; hence, larger population sizes are generally better for avoiding extinction. However, the effects of selection, migration and mutation are confounded with the sampling errors incurred in real populations of finite size. Thus, if populations are separated and selected for different environmental optima, migration can introduce alleles in high frequency that depress fitness. On the other hand, migration can maintain alleles in populations where they might be useful, but would be lost due to sampling accidents in small populations. It is reasonable to expect that all species are under the influence of several or all of the forces of evolution simultaneously and that the balance between the several forces is not the same for all genes in all populations. Furthermore, since the physical and biotic environments of forests are rarely at an equilibrium, the genetic and ecological dynamics of most species would have to be considered to be in other than a stable state. The actual distribution of the multiple functional forms of alleles at each of the several tens of thousands of gene loci in each of the individuals of a population must therefore be the resultant of the mixture of forces felt at each locus, subject to historical events. The vast possibilities of mixed distributions are overwhelming, although it is also obvious that for most of the species we can study, species are not random mixtures of an infinite number of possible combinations.

GENE SELECTION AND MICROARRAY DATA CLASSIFICATION IN FOREST

A random forest method has been selected to perform both gene selection and classification of the microarray data. In this embedded method, the selection of smallest possible sets of genes with lowest error rates is the key factor in achieving highest classification accuracy.

Hence, improved gene selection method using random forest has been proposed to obtain the smallest subset of genes as well as biggest subset of genes prior to classification. The option for biggest subset selection is done to assist researchers who intend to use the informative genes for further research. Enhanced random forest gene selection has performed better in terms of selecting the smallest subset as well as biggest subset of informative genes with lowest out of bag error rates through gene selection. Furthermore, the classification performed on the selected subset of genes using random forest has lead to lower prediction error rates compared to existing method and other similar available methods.

Through various biological experiments conducted worldwide, large datasets of information has been increasing rapidly and more analysis is conducted each day to sort out the puzzle. Since there are many separate methods available for performing gene selection as well as classification, finding similar approach for both, has been of interest to many researchers. Gene selection focuses at identifying a small subset of informative genes from the initial data in order to obtain high predictive accuracy for classification. Gene selection can be considered as a combinatorial search problem and therefore can be suitably handled with optimization methods.

Besides that, gene selection plays an important role preceding to tissue classification, as only important and related genes are selected for the classification. The main reason to perform gene selection is to identify a small subset of informative genes from the initial data before classification in order to obtain higher prediction accuracy. Many researchers use single variable rankings of the gene relevance and random thresholds to select the number of genes, which can only be applied to two class problems. Random forest can be used for problems arising from more than two classes (multi class) as stated by Díaz-Uriarte R & Alvarez de Andrés (2006).

Classification is carried out to correctly classify the testing samples according to the class. Therefore, performing gene selection antecedent to classification would severely improve the prediction accuracy of the microarray data. Random forest is an ensemble classifier which uses recursive partitioning to generate many trees and then combine the result. Using a bagging technique first proposed by Breiman (1996), each tree is independently constructed using a bootstrap sample of the data. Classification generates gene expression profiles which can discriminate between different known cell types or conditions as described by Lee *et al.* (2004). A classification problem is said to be binary in the event when there are only two class labels present and a classification problem is said to be a multiclass classification problem if there are at least three class labels. An enhanced version of gene selection using random forest is proposed to improve the gene selection as well as classification in order to achieve higher prediction accuracy.

The proposed idea is to select the smallest subset of genes with the lowest out of bag (OOB) error rates for classification. Besides that, the selection of biggest subset of genes with the lowest OOB error rates is also available to further improve the classification accuracy. Both options are provided as the gene selection technique is designed to suit the clinical or research application and it is not restricted to any particular microarray dataset. Apart from that, the option for setting the minimum number of genes to be selected is added to further improve the functionality of the gene selection method. Therefore, the minimum number of genes required can be set for gene selection process.

METHODOLOGY

Few improvements have been made to the existing random forest gene selection, which includes automated dataset input that simplifies the task of loading and processing of the dataset to an appropriate format so that it can be used in this software. Furthermore, the gene selection technique is improved by focusing on smallest subset of genes while taking into account lowest OOB error rates as well as biggest subset of genes with lowest OOB error rates that could increase the prediction accuracy. Besides that, additional functionalities are added to suite different research outcome and clinical application such as the range of the minimum required genes to be selected as a subset. Integration of the different approaches into a single function with parameters as an option allows greater usability while maintaining the computation time required.

Selection of Smallest Subset of Genes with Lowest OOB Error Rates

The existing method performs gene selection based on random forest to select smallest subset of genes while compromising on the out of bag (OOB) error rates. The subset of genes is usually small but the OOB error rates are not the lowest out of all the possible selection through backward elimination.

Therefore, enhancement has been made to improve the prediction accuracy by selecting the smallest subset with the lowest OOB error rates. Hence, lower prediction error rates can be achieved for classification of the samples. This technique is implemented in the random forest gene selection method and shown in Figure (see supplementary material) under supplementary section. During each subset selection based on backward elimination, the mean OOB error rate and standard deviation OOB error rate are tracked at every loop as the less informative genes are removed gradually. Once the loop terminates the subset with the smallest number of variables and lowest OOB error rates are selected for classification. The subset of genes is located based on the last iteration with the smallest OOB error rates. During the backward elimination process, the number of selected variables decreases as the iteration increases.

Selection of Biggest Subset of Genes with Lowest OOB Error Rates

Another method for improving the prediction error rates is by selecting the biggest subset with the lowest OOB error rates. This is due to the fact that any two or more subsets with different number of selected variables with same lowest error rates indicates that the informative genes level are the same, but the contribution of each genes towards the prediction accuracy is not the same. So, having more informative genes can increase the classification accuracy of the sample.

The technique applied for the selection of biggest subset of genes with the lowest OOB error rates are similar to the smallest subset of genes with the lowest OOB error rates, except that the selection is done by picking the first subset with the lowest OOB error rates from all the selected subset which has the lowest error rates. If there is more than one subset with lowest OOB error rates, the selection of the subset is done by selecting the one with highest number of variables for this method.

The detailed process flow for this method can be seen in the Figure (see supplementary material). This technique is implemented to assist researches that require filtration of genes for reducing the size of microarray dataset while making sure that the numbers of informative genes are high. This is achieved by eliminating unwanted genes as low as possible while achieving highest accuracy in prediction. Further enhancement is made to the existing random forest gene selection process by adding an extra functionality for specifying the minimum number of genes to be selected in the gene selection process that is included into the classification of the samples. This option allows flexibility of the program to suite the clinical research requirements as well as other application requirement based on the number of genes needed to be considered for classification. The selected minimum values are used during the backward elimination process which takes place in determining the best subset of genes based on out of bag (OOB) error rates.

Performance Measurement

For gene selection using random forest, backward elimination using OOB error rates is used as the final set of genes is selected based on the lowest out of bag (OOB) error rates as random forest returns a measure of error rate based on the out-of-bag cases for each fitted tree. The classification performance of the microarray data using random forest is measured using.632 bootstrap methods. In this method, the prediction error rates obtained is used to compare the performance of the random forest in classification where lower error rates means higher prediction accuracy. In the .632 bootstrap, accuracy is estimated as followed. Given a dataset of size n, a bootstrap sample is created by sampling n instances uniformly from the data (with replacement). Since the dataset is sampled with replacement, the probability of any given instance not being chosen after n samples is given in the supplementary material.

The expected number of distinct instances from the original dataset appearing in the test set is thus 0.632. The accuracy estimate is derived by using the bootstrap sample for training and the rest of the instances for testing. Given a number b, the number of bootstrap samples, let $c0_i$ be the accuracy estimate for bootstrap sample i. The .632 bootstrap estimates are defined as given in the supplementary material. The assessment method used has been able to populate and list the overall performance of the algorithm with other similar algorithms and techniques through prediction error rates calculation comparison.

Results & Discussion

In this section, the full result of all the options used is compared. In Figure, the result for each dataset is plotted against the accuracy, therefore the higher the values the lower is the error rates. Based on the Figure, the enhanced random forest gene selection performs better compared to standard method. Though, different options have different effects to the datasets being tested. Most of the datasets tested showed larger improvement in terms of accuracy achieved for classification when the subset of genes selected is larger. The detailed information regarding the datasets has been tabulated in Table (see supplementary material). However, some datasets with smaller subset of genes outperformed the larger subset of genes. This could be due to the effect of the informative genes, as more informative genes contribute to better classification accuracy. For the Leukemia dataset, either the selection of biggest subset of genes or limiting the range of the minimum number of genes to be selected in a particular subset has reduced the prediction accuracy. This is due to the fact that low number of informative genes contributes less to the overall classification accuracy. The highest accuracy achieved for this dataset is by selecting smallest

GENE SELECTION AND CLASSIFICATION OF MICROARRAY DATA USING RANDOM FOREST

Selection of relevant genes for sample classification (e.g., to differentiate between patients with and without cancer) is a common task in most gene expression studies e.g.. When facing gene selection problems, biomedical researchers often show interest in one of the following objectives:

1. To identify relevant genes for subsequent research; this involves obtaining a (probably large) set of genes that are related to the outcome of interest, and this set should include genes even if they perform similar functions and are highly correlated.
2. To identify small sets of genes that could be used for diagnostic purposes in clinical practice; this involves obtaining the smallest possible set of genes that can still achieve good predictive performance (thus, "redundant" genes should not be selected).

We will focus here on the second objective. Most gene selection approaches in class prediction problems combine ranking genes (e.g., using an F-ratio or a Wilcoxon statistic) with a specific classifier (e.g., discriminant analysis, nearest neighbor). Selecting an optimal number of features to use for classification is a complicated task, although some preliminary guidelines, based on simulation studies by [4], are available. Frequently an arbitrary decision as to the number of genes to retain is made e.g., keep the 50 best ranked genes and use them with a linear discriminant analysis as in; keep the best 150 genes as in. This approach, although it can be appropriate when the only objective is to classify samples, is not the most appropriate if the objective is to obtain the smaller possible sets of genes that will allow good predictive performance. Another common approach, with many variants e.g., is to repeatedly apply the same classifier over progressively smaller sets of genes (where we exclude genes based either on the ranking statistic or on the effect of the elimination of a gene on error rate) until a satisfactory solution is achieved (often the smallest error rate over all sets of genes tried).

A potential problem of this second approach, if the elimination is based on univariate rankings, is that the ranking of a gene is computed in isolation from all other genes, or at most in combinations of pairs of genes, and without any direct relation to the classification algorithm that will later be used to obtain the class predictions. Finally, the problem of gene selection is generally regarded as much more problematic in multi-class situations (where there are three or more classes to be differentiated), as evidence by recent papers in this area e.g.. Therefore, classification algorithms that directly provide measures of variable importance (related to the relevance of the variable in the classification) are of great interest for gene selection, specially if the classification algorithm itself presents features that make it well suited for the types of problems frequently faced with microarray data. Random forest is one such algorithm.

Random forest is an algorithm for classification developed by Leo Breiman that uses an ensemble of classification trees [14-16]. Each of the classification trees is built using a bootstrap sample of the data, and at each split the candidate set of variables is a random subset of the variables. Thus, random forest uses both bagging (bootstrap aggregation), a successful approach for combining unstable learners [16,17], and random variable selection for tree building. Each tree is unpruned (grown fully), so as to obtain low-bias trees; at the same time, bagging and random variable selection result in low correlation of the individual trees. The algorithm yields an ensemble that can achieve both low bias and low variance (from averaging over a large ensemble of low-bias, high-variance but low correlation trees). Random forest has excellent performance in classification tasks, comparable to support vector machines. Although random forest is not widely used in the microarray literature, it has several characteristics that make it ideal for these data sets:

- Can be used when there are many more variables than observations.
- Can be used both for two-class and multi-class problems of more than two classes.
- Has good predictive performance even when most predictive variables are noise, and therefore it does not require a pre-selection of genes (i.e., "shows strong robustness with respect to large feature sets",*sensu*).
- Does not overfit.
- Can handle a mixture of categorical and continuous predictors.
- Incorporates interactions among predictor variables.
- The output is invariant to monotone transformations of the predictors.
- There are high quality and free implementations: the original Fortran code from L. Breiman and A. Cutler, and an R package from A. Liaw and M. Wiener.
- Returns measures of variable (gene) importance.
- There is little need to fine-tune parameters to achieve excellent performance. The most important parameter to choose is *mtry*, the number of input variables tried at each split, but it has been reported that the default value is often a good choice. In addition, the user needs to decide how many trees to grow for each forest (*ntree*) as well as the minimum size of the terminal nodes (*nodesize*). These three parameters will be thoroughly examined in this paper.

Given these promising features, it is important to understand the performance of random forest compared to alternative state-of-the-art prediction methods with microarray data, as well as the effects of changes in the parameters of random forest. In this paper we present, as necessary background for the main topic of the paper (gene selection), the first through examination of these issues, including evaluating the effects of *mtry*, *ntree* and *nodesize* on error rate using nine real microarray data sets and simulated data. The main question addressed in this paper is gene selection using random forest. A few authors have previously used variable selection with random forest and use filtering approaches and, thus, do not take advantage of the measures of variable importance returned by random forest as part of the algorithm. Svetnik, Liaw, Tong and Wang propose a method that is somewhat similar to our approach. The main difference is that first find the "best" dimension (p) of the model, and then choose the p most important variables. This is a sound strategy when the objective is to build accurate predictors, without any regards for model interpretability. But this might not be the most appropriate for our purposes as it shifts the emphasis away from selection of specific genes, and in genomic studies the identity of the selected genes is relevant (e.g., to understand molecular pathways or to find targets for drug development).

The last issue addressed in this paper is the multiplicity (or lack of uniqueness or lack of stability) problem. Variable selection with microarray data can lead to many solutions that are equally good from the point of view of prediction rates, but that share few common genes. This multiplicity problem has been emphasized by and and recent examples are shown in Although multiplicity of results is not a problem when the only objective of our method is prediction, it casts serious doubts on the biological interpretability of the results. Unfortunately most "methods papers" in bioinformatics do not evaluate the stability of the results obtained, leading to a false sense of trust on the biological interpretability of the output obtained. Our paper presents a through and critical evaluation of the stability of the lists of selected genes with the proposed (and two competing) methods.

In this paper we present the first comprehensive evaluation of random forest for classification problems with microarray data, including an assessment of the effects of changes in its parameters and we show it to be an excellent performer even in multi-class problems, and without any need to fine-tune parameters or pre-select relevant genes. We then propose a new method for gene selection in classification problems (for both two-class and multi-class problems) that uses random forest; the main advantage of this method is that it returns very small sets of genes that retain a high predictive accuracy, and is competitive with existing methods of gene selection.

EVALUATION OF PERFORMANCE AND COMPARISONS WITH ALTERNATIVE APPROACHES

We have used both simulated and real microarray data sets to evaluate the variable selection procedure. For the real data sets, original reference paper and main features and further details are provided in the supplementary material. To evaluate if the proposed procedure can recover the signal in the data and can eliminate redundant genes, we need to use simulated data, so that we know exactly which genes are relevant. Details on the simulated data are provided in the methods and in the supplementary material.

Estimation of error rates

To estimate the prediction error rate of all methods we have used the .632+ bootstrap method. The .632+ bootstrap method uses a weighted average of the resubstitution error (the error when a classifier is applied to the training data) and the error on samples not used to train the predictor (the "leave-one-out" bootstrap error); this average is weighted by a quantity that reflects the amount of overfitting. It must be emphasized that the error rate used when performing variable selection is not what we report in as prediction error rate. To calculate the prediction error rate as reported, for example, the .632+ bootstrap method is applied to the complete procedure, and thus the samples used to compute the leave-one-out bootstrap error used in the .632+ method are samples that are not used when fitting the random forest, or carrying out variable selection. The .632+ bootstrap method was also used when evaluating the competing methods. Stability of variable (gene) selection evaluated using 200 bootstrap samples. "# Genes": number of genes selected on the original data set. "# Genes boot.": median (1st quartile, 3rd quartile) of number of genes selected from on the bootstrap samples....

Effects of parameters of random forest on prediction error rate

Before examining gene selection, we first evaluated the effect of changes in parameters of random forest on its classification performance. Random forest returns a measure of error rate based on the out-of-bag cases for each fitted tree, the OOB error, and this is the measure of error we will use here to assess the effects of parameters. We examined whether the OOB error rate is substantially affected by changes in*mtry*, *ntree*, and *nodesize*.

For both real and simulated data, the relation of OOB error rate with *mtry* is largely independent of *ntree* (for *ntree* between 1000 and 40000) and *nodesize* (nodesizes 1 and 5). In addition, the default setting of *mtry* (*mtryFactor* = 1 in the figures) is often a good choice in terms of OOB error rate. In some cases, increasing *mtry* can lead to small decreases in error rate, and decreases in *mtry* often lead to increases in the error rate. This is specially the case with simulated data with very few relevant genes (with very few relevant genes, small *mtry* results in many trees being built that do not incorporate any of the relevant genes). Since the OOB error and the relation between OOB error and *mtry* do not change whether we use *nodesize* of 1 or 5, and because the increase in computing speed from using *nodesize* of 5 is inconsequential, all further analyses will use only the default *nodesize* = 1. These results show the robustness of random forest to changes in its parameters; nevertheless, to re-examine robustness of gene selection to these parameters, in the rest of the paper we will report results for different settings of *ntree* and *mtry* (and these results will again show the robustness of the gene selection results to changes in *ntree* and *mtry*).

Gene selection using random forest

Random forest returns several measures of variable importance. The most reliable measure is based on the decrease of classification accuracy when values of a variable in a node of a tree are permuted randomly, and this is the measure of variable importance that we will use in the rest of the paper. (In the Supplementary material we show that this measure of variable importance is not the same as a non-parametric statistic of difference between groups, such as could be obtained with a Kruskal-Wallis test). Other measures of variable importance are available, however, and future research should compare the performance of different measures of importance.

To select genes we iteratively fit random forests, at each iteration building a new forest after discarding those variables (genes) with the smallest variable importances; the selected set of genes is the one that yields the smallest OOB error rate. Note that in this section we are using OOB error to choose the final set of genes, not to obtain unbiased estimates of the error rate of this rule. Because of the iterative approach, the OOB error is biased down and cannot be used to asses the overall error rate of the approach, for reasons analogous to those leading to "selection bias". To assess prediction error rates we will use the bootstrap, not OOB error. In our algorithm we examine all forests that result from eliminating, iteratively, a fraction,*fraction.dropped*, of the genes (the least important ones) used in the previous iteration.

By default,*fraction.dropped* = 0.2 which allows for relatively fast operation, is coherent with the idea of an "aggressive variable selection" approach, and increases the resolution as the number of genes considered becomes smaller. We do not recalculate variable importances at each step as mention severe overfitting resulting from recalculating variable importances. After fitting all forests, we examine the OOB error rates from all the fitted random forests. We choose the solution with the smallest number of genes whose error rate is within u standard errors of the minimum error rate of all forests. Setting $u = 0$ is the same as selecting the set of genes that leads to the smallest error rate. Setting $u = 1$ is similar to the common "1 s.e. rule", used in the classification trees literature [14,15]; this strategy can lead to solutions with fewer genes than selecting the solution with the smallest error rate, while achieving an error rate that is not different, within sampling error, from the "best solution". In this paper we will examine both the "1 s.e. rule" and the "0 s.e. rule".

On the simulated data sets backwards elimination often leads to very small sets of genes, often much smaller than the set of "true genes". The error rate of the variable selection procedure, estimated using the .632+ bootstrap method, indicates that the variable selection procedure does not lead to overfitting, and can achieve the objective of aggressively reducing the set of selected genes. In contrast, when the simplification procedure is applied to simulated data sets without signal (see Tables Tables11 and and22 in Additional file 1), the number of genes selected is consistently much larger and, as should be the case, the estimated error rate using the bootstrap corresponds to that achieved by always betting on the most probable class.

For additional results using different combinations of *ntree* = {2000, 5000, 20000}, *mtryFactor* = {1, 13}, *se*= {0, 1}, *fraction.dropped* = {0.2, 0.5}). Error rates when performing variable selection are in most cases comparable (within sampling error) to those from random forest without variable selection, and comparable also to the error rates from competing state-of-the-art prediction methods. The number of genes selected varies by data set, but generally the variable selection procedure leads to small (< 50) sets of predictor genes, often much smaller than those from competing approaches. There are no relevant differences in error rate related to differences in *mtry*, *ntree* or whether we use the "s.e. 1" or "s.e. 0" rules. The use of the "s.e. 1" rule, however, tends to result in smaller sets of selected genes.

Stability (uniqueness) of results

We have evaluated the stability of the variable selection procedure using the bootstrap. This allows us to asses how often a given gene, selected when running the variable selection procedure in the original sample, is selected when running the procedure on bootstrap samples. The results here will focus on the real microarray data sets. For other combinations of*ntree*, *mtryFactor*, *fraction.dropped*, *se*) shows the variation in the number of genes selected in bootstrap samples, and the frequency with which the genes selected in the original sample appear among the genes selected from the bootstrap samples. In most cases, there is a wide range in the number of genes selected; more importantly, the genes selected in the original samples are rarely selected in more than 50% of the bootstrap samples. These results are not strongly affected by variations in *ntree* or *mtry*; using the "s.e. 1" rule can lead, in some cases, to increased stability of the results.

Discussion

We have first presented an exhaustive evaluation of the performance of random forest for classification problems with microarray data, and shown it to be competitive with alternative methods, without requiring any fine-tuning of parameters or pre-selection of variables. The performance of random forest without variable selection is also equivalent to that of alternative approaches that fine-tune the variable selection process. We have then examined the performance of an approach for gene selection using random forest, and compared it to alternative approaches. Our results, using both simulated and real microarray data sets, show that this method of gene selection accomplishes the proposed objectives. Our method returns very small sets of genes compared to alternative variable selection methods, while retaining predictive performance. Our method of gene selection will not return sets of genes that are highly correlated, because they are redundant. This method will be most useful under two scenarios: a) when considering the design of diagnostic tools, where having a small set of probes is often desirable; b) to help understand the results from other gene selection approaches that return many genes, so as to understand which ones of those genes have the largest signal to noise ratio and could be used as surrogates for complex processes involving many correlated genes. A backwards elimination method, precursor to the one used here, has been already used to predict breast tumor type based on chromosomic alterations.

We have also thoroughly examined the effects of changes in the parameters of random forest (specifically *mtry*, *ntree*, *nodesize*) and the variable selection algorithm (*se*,*fraction.dropped*). Changes in these parameters have in most cases negligible effects, suggesting that the default values are often good options, but we can make some general recommendations. Time of execution of the code increases H" linearly with *ntree*. Larger *ntree* values lead to slightly more stable values of variable importances, but for the data sets examined, *ntree* = 2000 or *ntree* = 5000 seem quite adequate, with further increases having negligible effects. The change in *nodesize* from 1 to 5 has negligible effects, and thus its default setting of 1 is appropriate. For the backwards elimination algorithm, the parameter *fraction.dropped* can be adjusted to modify the resolution of the number of variable selected; smaller values of*fraction.dropped* lead to finer resolution in the examination of number of genes, but to slower execution of the code. Finally, the parameter *se* has also minor effects on the results of the backwards variable selection algorithm but a value of *se* = 1 leads to slightly more stable results and smaller sets of selected genes.

In contrast to other procedures our procedure does not require to pre-specify the number of genes to be used, but rather adaptively chooses the number of genes have conducted an evaluation of several gene selection algorithms, including genetic algorithms and various ranking methods; these authors show results for the Leukemia and NCI60 data sets, but the Leukemia results are not directly comparable since focus on a three-class problem. They report the best results with the NCI60 data set estimated with the .632 bootstrap rule. These best error rates are 0.408 for their evolutionary algorithm with 30 genes and 0.318 for 40 top-ranked genes. Using a number of genes slightly larger than us, these error rates are similar to ours; however, these are the best error rates achieved over a range of ranking methods and error rates, and not the result of a complete procedure that automatically determines the best number of genes and ranking scheme. A comparative study of feature selection and multi-class classification. Although they use four-fold cross-validation instead of the bootstrap to assess error rates, their results for three data sets common to both studies (Srbct, Lymphoma, NCI60) are similar to, or worse than, ours. In contrast to our method, their approach pre-selects a set of 150 genes for prediction and their best error rates are those over a set of seven different algorithms and eight different rank selection methods, where no algorithm or gene selection was consistently the best. In contrast, our results with one single algorithm and gene selection method (random forest) match or outperform their results.

Recently, several approaches that adaptively select the best number of genes or features have been reported. For the Leukemia data set our method consistently returns sets of two genes, similar to using an exhaustive search method, and lower than the numbers given by of 3 to 25 have proposed a Bayesian model averaging (BMA) approach for gene selection; comparing the results for the two common data sets between our study and theirs, in one case (Leukemia) our procedure returns a much smaller set of genes (2 vs. 15), whereas in another (Breast, 2 class) their BMA procedure returns 8 fewer genes (14 vs. 6); in contrast to BMA, however, our procedure does not require setting a limit in the maximum number of relevant genes to be selected have developed a method for gene selection and classification, LS Bound, related to least-squares SVMs; their method uses an initial pre-filtering (they choose 1000 initial genes) and is not clear how it could be applied to multi-class problems.

The performance of their procedure with the leukemia data set is better than that reported by our method, but they use a total of 72 samples (the original 38 training plus the 34 validation of [44]) thus making these results hard to compare. With the colon data sets, however, their best performing results are not better than ours with a number of features that is similar to ours. Proposed two Bayesian classification algorithms that incorporate gene selection (though it is not clear how their algorithms can be used in multi-class problems).

The results for the Leukemia data set are not comparable to ours (as they use the validation set of 34 samples), but their results for the colon data set show error rates of 0.167 to 0.242, slightly larger than ours (although these authors used random partitions with 50 training and 12 testing samples instead of the .632+ bootstrap to assess error rate), with between 8 and 15 features selected (somewhat larger than those from random forest). Finally, [31], applied both shrunken centroids and a genetic algorithm + KNN technique to the NCI60 and Srcbt data sets; their results with shrunken centroids are similar to ours with that technique, but the genetic algorithm + KNN technique used larger sets of genes (155 and 72 for the NCI60 and Srbct, respectively) than variable selection with random forest using the suggested parameters. In summary, then, our proposed procedure matches or outperforms alternative approaches for gene selection in terms of error rate and number of genes selected, without any need to fine-tune parameters or preselect genes; in addition, this method is equally applicable to two-class and multi-class problems, and has software readily available. Thus, the newly proposed method is an ideal candidate for gene selection in classification problems with microarray data.

A reviewer has alerted us to the paper by Jiang et al., previously unknown to us. In fact, our approach is virtually the same as the one used by Jiang et al., with the exception that these authors recompute variable importances at each step (we do not do this in this paper, although the option is available in our code) and, more importantly, that their gene selection is based both in the OOB error, as well as the prediction error when the forest trained with one data set is applied to a second, independent, data set; thus, this approach for gene selection is not feasible when we only have one data set. Jiang et al. [45] also show the excellent performance of variable selection using random forest when applied to their data sets. The final issue addressed in this paper is instability or multiplicity of the selected sets of genes. From this point of view, the results are slightly disappointing. But so are the results of the competing methods.

And so are the results of most examined methods so far with microarray data, as shown in [29] and [30] and discussed thoroughly by [27] for classification and by [28] for the related problem of the effect of threshold choice in gene selection. However, and except for the above cited papers and [6,46] and [5], this is an issue that still seems largely ignored in the microarray literature. As these papers and the statistical literature on variable selection (e.g., [40,47]) discusses, the causes of the problem are small sample sizes and the extremely small ratio of samples to variables (i.e., number of arrays to number of genes). Thus, we might need to learn to live with the problem, and try to assess the stability and robustness of our results by using a variety of gene selection features, and examining whether there is a subset of features that tends to be repeatedly selected. This concern is explicitly taken into account in our results, and facilities for examining this problem are part of our R code.

The multiplicity problem, however, does not need to result in large prediction errors. That the very different classifiers often lead to comparable and successful error rates with a variety of microarray data sets. Thus, although improving prediction rates is important, when trying to address questions of biological mechanism or discover therapeutic targets, probably a more challenging and relevant issue is to identify sets of genes with biological relevance. Two areas of future research are using random forest for the selection of potentially large sets of genes that include correlated genes, and improving the computational efficiency of these approaches; in the present work, we have used parallelization of the "embarrassingly parallelizable" tasks using MPI with the Rmpi and Snow packages [50,51] for R. In a broader context, further work is warranted on the stability properties and biological relevance of this and other gene-selection approaches, because the multiplicity problem casts doubts on the biological interpretability of most results based on a single run of one gene-selection approach.

2

Conservation of Forest Genetic Resources and Breeding

Although the goal of conserving forest genetic resources can be simply stated, its implementation can be very complex and expensive. With thousands of tree species distributed among several local populations (interbreeding groups of individuals), each with thousands of variable genetic loci, priorities should be set first at the species level; only then can we assign priorities among populations. It is important that data on species that are of significance to conserve, and the levels of threat to them, be collated with in situ and ex situ management approaches in mind. In practical terms, national programmes need ways of establishing priorities for conservation that take into account the large potential number of species for which they may be responsible. Sometimes the focus may be on species, because of their charismatic appeal. Alternatively, the focus may have to be on perceived threats resulting from economic values or ecological traits: for example, species that have low population densities, highly specialized pollination patterns or particular seed germination mechanisms. Baseline information on the status of genetic diversity, a rating of species potential value, an evaluation of the threats and the potential for conservation management are some of the necessary steps for priority-setting. The outcome can be a ranking of priorities for management or a classification of species into priority groups.

In general, an effective species conservation programme needs to take into account the whole range of geographic distribution of a species, as well as the species metapopulation structure. Without this information, one cannot claim that the genetic diversity of the target species is conserved overall. Most national conservation programmes for forest genetic resources must therefore deal with the conservation of locally adapted populations. In an ideal case, the geographic distribution of a species would be listed and mapped, the type and extent of threats to particular populations would be known, and methods of conservation and management would be well established. It would then be possible to evaluate the impacts of different threats, the costs and the effectiveness of different management options in minimizing the associated impacts. Priorities could then be established, based on the evaluation of those resources for economic or ecological payoffs. However, we rarely have such complete or detailed knowledge. Many tree species that are known to be under threat of extinction are not included in conservation programmes. In some instances, the genetic resources may be well conserved within protected areas, but these might represent just a small fraction of the overall genetic variation of the species. Patterns of genetic variation may often be cryptic and reductions may portend demographic collapse as either a causal or an associated agent. Therefore, demographic data, if available, is not sufficient, nor are listings of species in parks sufficient for planning conservation. However, collating such data and testing for genetic patterns of variation can provide at least initial indicators of the state of the genetic resource.

Targeting species and regions for more detailed genetic surveys can then provide information on priority species for conservation. These authors also suggest two levels of targeting; one for research and risk assessments, and the second for management of specific risks. Large-scale data, such as are available from geographic information systems (GIS) analyses, can provide some indications of demographic threats, and hence multiple sources of data to be collated and analysed simultaneously. In the first targeting efforts of the former IBPGR (International Board of Plant Genetic Resources, currently International Plant Genetic Resources Institute, IPGRI), the priority lists of crop species that guided collection efforts were created largely on the basis of likelihoods of demographic extinction and thus on risks of genetic erosion, made possible by a broad global sharing of common objectives and information from crop species in agricultural systems.

Unfortunately, there is no such consensus for forest trees and hence more explicit statements of models and factors are needed at an international level, as well as for any country trying to establish its own priorities. In most conservation programmes, people must make decisions about management actions, even with limited access to research data to support their decisions. In these cases, they must also decide which characteristics of the organisms involved are least known but, if studied, are most likely to make a difference in how the organisms are managed. In this sense, we are interested not so much in descriptive biology as in the biology of genetically and ecologically critical functions, such as adaptation to changing environments and potential for evolution and artificial selection.

CONVERSION OF FORESTS TO OTHER LAND USES

Forests can be lost either because forest resources and trees are not regarded as being of economic importance, or because of a policy framework that makes it possible to replace forests with other land uses. Often this is based on short-term maximization of economic returns and lack of supportive forest policies based on good understanding of the potential of forests as sources of income and products for local and regional markets and their associated services for other sectors of the economy.

CONSERVE: ECOSYSTEMS, SPECIES OR GENES

The first step in genetic conservation is to specify the objectives of the conservation programme. This is of the utmost importance, since it is possible to conserve ecosystem properties and still lose species entities. It is also possible to conserve a species and still lose genetically distinct populations, and therefore, genes that may be of value in disease and pest resistance, and in future adaptation. They could also be important in species deployment through breeding programmes, if that becomes a need or necessity. Many forest genera and species around the world provide goods and services, such as timber, wood, food, fodder, environmental stabilization, shade, shelter, and cultural and spiritual values. However, fewer than 1 000 tree species have been systematically tested for their present-day utility, and less than 100 are the subject of intensive genetic research programmes. Evidently, therefore, many different forest species are being used in situ to provide important goods and services, without any active genetic management.

Worldwide, the conservation of forest genetic resources has as its overall objective the maintenance of genetic diversity in the thousands of tree species of known or potential socioeconomic and environmental importance. Moreover, the levels and distribution of genetic variation in any given species are expected to be in a process of constant natural change resulting from the main forces of evolution. Therefore, the central concern of conservation should be the evolutionary processes which promote and maintain genetic diversity, and not the endeavour to preserve the present distribution of variation as an end in itself.

Assessing Species' Priorities for Conservation Action

Within any given country or local area there may be divergent opinions on priorities among tree species. Forestry departments are likely to have a somewhat different emphasis and priorities from those of local forest dwellers and users, which can be different again from those of farmers and various other users of trees. It is apparent that in situ conservation programmes will be more successful if they target species of direct interest, use or concern to the land management authority and/or landowner(s): this will have major implications for the planning of in situ conservation programmes. A participatory rural appraisal approach can be useful for helping local communities whose land is communally owned to better identify their priority tree genetic resources and to develop appropriate in situ conservation responses. It is also important to consider the case for species and provenances which are of major economic importance when planted as exotics, but currently of much less significance in their native range and habitats; an example is Pinus radiata from the south-western USA and Mexico. In such cases it is not unreasonable to expect that the likely beneficiaries of in situ conservation should contribute, financially or otherwise, to conservation. Given that there will be limited financial resources available for specific conservation programmes for forest genetic resources, it is necessary to consider which of the priority species are also in most need of, or warrant, conservation interventions and actions. This can be conveniently undertaken for different species by comparing the extent of the resource (level of genetic diversity or intraspecific variation) with the vulnerability or threats to the populations and/or ecosystems of which they are a part.

In Situ Conservation Atrategies

In the case of non-domesticated species, in situ conservation is probably the most important strategy and sometimes the only viable approach. In the tropics, where extinction rates of species are high because of land-use changes, setting conservation priorities is critical. This is particularly evident in developing countries, where resources allocated for conservation are scarce and baseline information on species distribution and richness data are lacking. In a world of scarce resources, one approach to priority-setting is through networking activities, with initiatives involving multiple countries and stakeholders. In situ conservation is usually the preferred conservation strategy for most wild plant species, including some of the wild relatives of crop species, because, it allows the populations of interest to continue to be exposed to evolutionary processes. Alternatively, for many domesticated species (crop and livestock), on-farm conservation of traditional varieties is now widely supported as an important practice for conservation of genetic diversity.

Molecular genetic studies, carried out on many forest tree species around the world, are contributing to a better understanding of patterns of variation to support the development of improved management practices, and to monitor changes of species turnover in time and in space. In some situations, these priorities could be refined by the use of new tools, such as molecular markers and modelling simulations. Integrating GIS tools with molecular research will improve our knowledge of landscape patterns of genetic diversity of species distribution, and help develop resource management plans. For example, in the Western Ghats, India, these two approaches are being used in combination to detect areas with high diversity (intra- and interspecific) and to set priorities for conservation. Molecular markers can thus assist in the conservation of tree species and may allow national programmes to reflect biodiversity patterns in their own management plans. Molecular methods can also help to identify differences between local and non-local provenances and genotypes, identify where diversity is being lost, and facilitate the introduction of new diversity for integrated conservation and breeding programmes. This can also provide for a better management of intraspecific genetic variation.

Evolutionary Conservation

Evolutionary conservation activities are characterized by programmes where the trees produce progeny in successive generations: genes are generally 'conserved', but genotypes are not. Natural selection takes place among trees with new allelic combinations that either favour or disfavour different genotypes. This process ensures that gene frequencies will change in the population: alleles with positive influence on fitness will increase, and alleles associated with low fitness will decrease. If the population size is sufficient, neutral genes should, in general, be maintained, but some genes will inevitably be lost by genetic drift; new genetic variation will arise by mutation after several generations. Human interventions (if any) are designed to facilitate moderate genetic processes rather than to avoid them. Genetic variation between populations is generally maintained when they are growing in different environments, and is even expected to increase over time.

A typical example of a conservation population capable of evolutionary processes is a protected area in a natural forest. In a protected area, the species occupies its natural habitat (it is said to be in situ conserved), typically with a wide range of other species. Natural selection for general fitness is therefore largely related to competition among species, as well as to adaptation within species to current and future environmental conditions. However, evolutionary conservation can also take place in a planted stand, if natural selection is allowed to work, and if the planted trees are regenerated from seed, rather than by vegetative techniques, for the next generation. In such programmes, plantations will preferably be established and managed in ways that mimic the natural processes that will support natural selection.

Of course, in most situations the mixture of species (if any) is largely artificial, and the selective forces may therefore favour different genes than would be the case in true in situ conservation. However, this reflects the fact that selection and fitness always depend on the degree of human influence in any ecosystem. Directional selection in favour of commercial traits—including characters such as good stem form or ease of establishment in plantations—is typically avoided in strict evolutionary conservation programmes, but of course this again depends upon the local objectives of the programme.

PLANT GENETIC RESOURCES

Effective conservation and use of plant genetic resources involve asking many questions about the extent and distribution of genetic variation. Only when the appropriate markers and technologies for describing this variation are accessible can such questions be adequately addressed.

The most appropriate markers for a given question should: (i) be heritable; (ii) discriminate between the individuals, populations, or taxa being examined; (iii) be easy to measure and evaluate; and (iv) provide results that can be compared with results of similar studies. Molecular markers and assays that meet these criteria were introduced in the 1950s and have proliferated ever since. Use of protein isozymes as genetic markers increased rapidly after their introduction. Nucleic acid markers gained popularity in the 1970s, with the advent of DNA sequencing and restriction fragment analysis. The number of different molecular markers and assays has increased exponentially since the late 1980s, when introduction of the polymerase chain reaction (PCR) made DNA amplification and sequencing possible and affordable.

As the diversity of markers and assays increases, the cost of using them continues to decrease. At the same time, the range of taxa and tissues that can be analysed has expanded. Improved DNA extraction, amplification, and sequencing protocols allow us to analyse ancient DNA, from 13,000-year-old specimens of *Mylodon* (ground sloths) and 17–20 million-year-old *Taxodium* (bald cypress) specimens, and samples as small as single cells or fractions of a seed. Thus the chances are good that at least one marker assay will be appropriate for any question about variation in any taxa. Furthermore, results of one study can generally be related to results from many other studies.

Such opportunities are inevitably accompanied by challenges. None of the challenges are unique to molecular studies, but all become more critical as the number of available marker assays increases. Faced with a bewildering variety of molecular technologies (and their acronyms), choosing a marker and assay that are appropriate for a scientist's particular question, taxa of interest, and practical constraints can be overwhelming. The temptation is great to use techniques that are recommended by coworkers, are frequently used, for which the most information is available, or are the newest – without adequately evaluating the techniques to determine which is most suitable. Because molecular markers can be measured in virtually all plant taxa and minimal tissue is required, extra care must be taken to verify that the entries in a test array are correctly identified. Since molecular markers are widely applicable and their constraints are not always understood, thoughtful planning is particularly important when designing molecular studies, to ensure that they have an appropriate range of entries, an appropriate number of entries per taxon, and conclusions that stay within the scope of a specific question and test array. In addition, the vast amount of published research using molecular markers complicates the task of thoughtfully relating one study to others with similar marker assays, research questions, and/or taxa.

Another challenge is formulating research objectives that ask critical questions about plant genetic resources. Descriptive studies that use molecular marker assays to examine variation between and within specific taxa can generate valuable information, but some plant conservation questions are best addressed by experimental studies. The wide variety of molecular marker assays available provides new ways to address such questions – if the markers are appropriately used. Such challenges can be met successfully and thoughtful decisions can be made if logical frameworks are available to categorize and compare these valuable molecular tools. A solid foundation for such comparisons has been provided recently. This chapter builds on that foundation by reviewing both early and recently developed molecular markers and assays, citing examples of their use. The types of questions asked about plant genetic resources are described, as are guidelines for selecting appropriate markers and assays to address these questions. The literature cited here is far from comprehensive, but includes applications of each marker assay in plant genetic research. While the list of molecular techniques will soon be dated, hopefully the framework presented will be helpful for categorizing and comparing new tools as they are made available. The chapter concludes with a perspective on new marker assays as they apply to, and may change, the study of plant genetic variation.

MARKERS AND MARKER ASSAYS

All studies of genetic variation have four primary components: (i) the research question; (ii) the marker used to address it; (iii) the technique used to generate and measure the marker; and (iv) the method used to analyse the data produced. This chapter focuses on the first three components.

PLANT GENETIC RESOURCE

Most questions asked about *ex situ* and *in situ* plant genetic variation fit into one of three categories. While boundaries between them are not always clear, these categories are useful when deciding how to address specific questions. Questions of identity are forensic in nature and ask whether two or more samples are genetically the same. These questions concern whether genebank accessions are present in duplicate, whether populations are genetically distinct or are geographical subdivisions of one population, and whether genetic change has occurred in an accession or population over time.

In contrast, questions of location and diagnostics concern the presence and location of a particular allele or nucleotide sequence in a taxon, genebank accession, *in situ* population, individual, chromosome, or cloned DNA segment. These questions are asked to locate populations or individuals that have desirable traits; determine whether traits have been lost from a population, or have been transferred from one population or plant to another; and establish the position of markers on physical or genetic maps. The differences between diagnostic and forensic questions are significant, but are not always clarified. Both types of question require qualitative information about specific loci. Yet while diagnostic studies often use single-locus markers, each with few alternative states, forensic questions are commonly addressed with information from many multiallelic loci.

Finally, relationship and structure questions examine relatedness or similarity between genotypes, the amount of genetic variation present, and how variation is distributed between individuals, populations, and taxa. Such questions are used to pinpoint gaps in genebank collections, decide which genotypes are high priority for *in situ* conservation, design germplasm sampling strategies and regeneration programmes, construct core collections, study gene flow, and determine optimum sizes for conserved populations or genebank accessions. These issues are generally addressed with information from many loci. Quantitative data can be used, but qualitative data are more appropriate for some phylogenetic and parentage questions.

Molecular Markers

The appropriate markers for a study can discriminate between entries in an array, but are not so polymorphic that important variation is masked by random noise. Molecular markers range from highly conserved to hypervariable, and can be either proteins or nucleic acids. The nucleic acids used as markers include entire genomes, single chromosomes, fragments of DNA or RNA, and single nucleotides.

A wide variety of nucleic acid fragments are utilized as markers. While some occur once in a genome, others are repeated. Many repeated sequences used as markers are noncoding; others are elements of multigene families. Some repeated sequences are interspersed throughout the genome, either distributed randomly or in clusters. These interspersed repeats are common in plant and animal nuclear genomes, and are found in plant (but not animal) mitochondrial genomes. The chloroplast genome contains a large inverted repeat (IR); most angiosperm chloroplasts have two copies, separated by a short single- copy region. Repeat length and (rarely) loss of one copy can vary between taxa. Much research at present is focused on repeated sequences that occur in tandem. The classes of tandem repeats are distinguished by the length of the core repeat unit, the number of repeat units per locus, and the abundance and distribution of loci. The names for these classes are themselves varied and have been inconsistently used, but Harding *et al.* (1992) and Tautz (1993) have clarified the nomenclature.

Tandem repeats were first reported in the literature as 'satellites' of DNA, detected in CsCl density gradients as fractions with different GC content than the rest of the genome. These satellites have repeat units that are usually several hundred nucleotides long, with thousands of copies at each of several loci in the nuclear genome. These loci are usually in heterochromatin, often near centromeres. Satellite DNA is present in numerous species. For many satellites, the number of loci and number of repeat units per locus vary between species and higher taxa. Minisatellites (often called variable number of tandem repeat loci, or VNTR loci) are widely used as markers, especially in forensics. The repeat units are usually less than 100 nucleotides long, with tens to hundreds of copies per locus. Thousands of loci in a genome may have similar core repeat units. The number of repeat units at a minisatellite locus can vary greatly between individuals and populations. First described in humans, minisatellites are found in numerous animal species, often near telomeres. They are also common in plants, and are often associated with satellites and centromeres. The number of repeat units per locus is less variable in plants than in animals, but is still high; plant minisatellites are useful markers for variation between and within species.

MOLECULAR MARKER ASSAYS

Molecular marker assays are generally classified by whether the molecules evaluated are proteins or nucleic acids, and whether the character analysed in a nucleic acid marker assay is the entire genome, a chromosome, a fragment, or a nucleotide. Alternatively, marker assays can be categorized by the type of character measured. Some methods measure quantitative differences between entries in an array. Others measure qualitative characters, each with two or more possible states. Marker assays also differ in the number of loci evaluated per analysis, whether multiple loci are evaluated simultaneously or sequentially, and the type and amount of information needed about the marker loci before conducting the assay.

Protein Marker Assays

When proteins are used as genetic markers, the assumption is made that any variation between proteins reflects heritable variation in their amino acid sequences. This assumption has long been debated, since protein phenotypes can be affected by factors such as post-translational modification, plant phenology, and environmental conditions during plant growth. For many tasks, however, protein marker assays continue to be useful molecular tools. Some assays measure immunological properties of proteins. When antigenic proteins of one plant or animal species are injected into an animal (usually a rabbit), antibodies are produced. Genetic differences between entries in an array can be assessed by comparing how each entry's proteins bind to antibody serum (antiserum) from a reference species. This binding affinity is measured by the amount of antibody–antigen precipitate produced or by microcomplement fixation (MCF). Each MCF solution contains antiserum from a reference sample, antigen from a test entry, and a protein complement. The complement can only bind to antibody that is bound to the antigen. The amount of unfixed complement is measured; this is used to calculate the amount of unbound antigen, which indicates the extent of variation (the immunological distance) between the entry and the reference sample. After MCF assays are conducted for each entry in an array, these immunological distances are compared between entries. MCF has long been used in vertebrate phylogenetic analysis.

Until recently, immunological markers were seldom used to study plant variation, perhaps because few suitable proteins had been identified. During the past few years, however, monoclonal antibodies to seed proteins of several cereal species have been produced. Reactivity of these antibodies with antigenic proteins can be evaluated by enzyme-linked immunosorbent assays (ELISAs). Monoclonal antibodies have been used to identify cDNA library clones and describe variation between crop species and cultivars. Other marker assays measure the rate of protein migration through a starch, polyacrylamide, or cellulose acetate gel in response to an electrical current. The migration rate is related to protein size, shape, isoelectric point and/or ionic charge.

In these assays, protein extracts from entries in an array are loaded in adjacent gel lanes and electrophoresed. The gel is then treated with a histochemical stain that reacts with a specific marker protein. Variation between the entries is measured by the positions of their stained proteins on the gel. Many of the proteins used as markers are isozymes – functionally similar forms of an enzyme, encoded by different loci or different alleles at a locus. In general all alleles at a locus are expressed, so allelic isozymes can be analysed as codominant variants.

Isozymes are frequently used to describe variation between and within plant populations, species, and genera; for mapping and diagnostics; and in some cases for identification. When comparing higher taxonomic levels, the probability of genetically different electromorphs migrating to the same gel position is increased. For closely related taxa and populations, however, isozyme analysis is informative and continues to be refined. Immunoelectrophoretic assays measure both protein migration rate in gels and the antibody specificity of proteins. Antigenic proteins are electrophoretically separated, then are detected by their reactions with antibodies. This technique is valuable for diagnostics, mapping, and describing variation between plant cultivars, species, and higher taxa.

GLOBAL SYSTEM ON CONSERVATION AND SUSTAINABLE USE OF PLANT GENETIC RESOURCES

The salient points related to the progress of the global system on conservation and sustainable use of plant genetic resources for food and agriculture which emerged in the seventh regular session of the FAO Commission on Genetic Resources for Food and Agriculture (CGRFA) at Rome during 15-23 May, 1997 may be reviewed as follows:

- A multi-year work programme on agricultural biodiversity established as per decision III/11 of COP shall be pursued by the FAO Commission and CBD Secretariat though cooperation. Similarly, FAO's monitoring/reporting on agricultural biodiversity shall be consistent with CBD, other relevant inter-governmental bodies and world food summit' Plan of Action.

- CBD through its decision II/5 has requested GEF to give priority support to efforts for conservation and sustainable use of agro-biodiversity. Many countries stressed the need to have separate funding! FAO commission requested FAO to provide assistance and guidance to countries on request in implementation/monitoring of the Global Plan of Action (GPA). Further, the monitoring of GPA and updating/reporting on the state of world's PGR were held complementary activities which do not need separate reports (*i.e.*, need only one report each) by countries.
- No changes in the CCGCT (Code of Conduct for Germplasm Collecting and Transfer) or any further work on Code of Conduct for Biotechnology were recommended, till the revision of IUPGR. However, a decision guide on regeneration of accessions in seed collections, prepared by IPGRI, has been made available to the FAO commission for comments by members.
- The existing arrangements between FAO and the 12 CGIAR centres shall be extended, pending the revision of IUPGR.
- Statutes of the inter-governmental technical working group on AGR (Animal Genetic Resources) and PGR were presented. Both working groups have 5 members each for Asia. The AGR were specifically addressed in view of the broadened mandate of the Commission and to facilitate integrated approach to agro-biodiversity.
- The summary statement presented by IPGRI informed the Commission that 0.5 million accession were now covered under the FAO-CGIAR agreements with the designation of 50,000 new accession under this arrangement.
- The summary statement presented by the CBD Secretariat showed that at least three decisions of the COP-III, Buenos Aires, Nov., 1996, were of direct relevance to the ongoing negotiation for revision of IUPGR, namely, Decision III/11-Conservation and Sustainable use of agro-biodiversity, Decision III/15-Access to Genetic Resources, and Decision III/17-Intellectual Property Rights. The COP-III noted the narrow options for legal status of revised International Undertaking, as being, i) a voluntary agreement, ii) a binding instrument, and iii) protocol to CBD. It encouraged parties to implement GPA at national levels. Pollinators and soil-micrograms in agriculture needed special attention (as bio-indicators) and comprehensive reporting by countries

to Subsidiary Body on Scientific, Technical and Technological Advice (SBSTTA)/Conference of Parties (COP) of the CBD.

- An identified group of competent people needs to resolve the PGR issues and help prioritise various PGR activities. There is need to identify crops/regions on priority.
- The dialogue between the public and private sector should continue and there is a need to create awareness about PGR.
- An information bank should be established to enhance literacy, management and utilisation of genetic wealth. Emphasis should also be given to regional languages in developing the information base.

World Information and Early Warning System

The FAO's initiative to develop a systematic information base on PGR and related technologies under the World Information and Early Warning System on Plant Genetic Resources (WIEWS) is aimed to collect, collate, maintain and disseminate facts and figures on country profiles, national PGR programmes, *ex situ* collections, database on PGR, existing crop varieties, seed production and supplying agencies, *in situ* on farm conservation, bioprospecting and exchange of seed. The nature of such an activity requires a comprehensive information input and a grid approach for analysis and interpretation of data. The system is open and responsive to queries from all stakeholders and interest groups. In practical terms, it would be relatively simpler to attend to details of the resource availability and the custodian(s) but, considering the importance of implementing CBD and WTO provisions and the imminent revision of the International Undertaking on Plant Genetic Resources, the access to resources for the purpose of replenishment could pose certain difficulties. The access would have to be required from any one of the following sources:

- The *ex situ* seed collections;
- The seed multiplication and supplying agencies; and
- The local communities/community gene banks.

The implications, in each case, could be several and manifold. The access from *ex situ* CG system collections is being debated wherein two clear-cut options have been put forth. The pre-CBD collections are still available freely which means their replenishment would be only a matter of time actually required for the multiplication of their seed. However, in case of requirement of post CBD collections, several access-related issues shall have to be sorted prior to the multiplication of seed for replenishment. The requirement of seed from commercial/public seed multiplication and supplying agencies could be easily met but such a replenishment would not serve the purpose effectively. This being so, the agencies could provide seed of few dominant varieties, which they are currently handling, in contrast with the requirement of a wider genetic base.

The replenishment of seed from farmers' varieties/land races faces two different situations:

- The issue of consent/access; and
- The issue of compensation or sharing of benefit accrued from the use of their heritage seed.

The setting up of community seed banks and the Global Biodiversity Fund have definitely been provided as suitable institutions which could help in build up at an appropriate system and come to the rescue in cases of crisis. The setting up of a separate Global Biodiversity Fund is still being debated as also the Farmers' Rights. The development of the *sui generis* systems to meet the countries' own (local) needs, has to be accelerated. Through this mechanism, the whole idea is to encourage local level replenishment through procurement of required seed from nearly/similar climate logical areas.

Some Queries Seeking Answers

Increased investments on *ex situ* conservation are unavoidable with each increment in the quantum of work and management skills. The cost/benefit ratio does not work in such cases because the conservation is for posterity. In this endeavour, the obvious questions seeking answers would be:

- What should be the estimate of funds for genetic resources conservation at the CG Centres, at the National genebanks and at community seed banks?
- What kind of activities we need to take up in the beginning which ensure fast replenishment in cases of crisis?
- What shall be the mechanism of funding-whether the Global Biodiversity Fund or private sector contribution shall be forthcoming or not?
- Which sites for *in situ*/on-farm conservation to begin with?
- How to identify genuine NGOs for their involvement in PGR Conservation?
- What could be a standard format of collecting and collating information for WIEWS on PGR *ex situ* conservation, country status, on-farm conservation and wild relatives? Do the existing proforma suffice?
- What would be a suitable indicator or an appropriate method for monitoring of genetic erosion for use under Early Warning Systems?
- What should be the suitable proforma for multicrop passport data?
- How to generate compatible country reports providing information for the State of the World's PGR and the action on GPA?
- How to minimise diversity and redundancy in use of computer software for various Database Management Activities?

GENETIC RESOURCES AND PLANT BREEDING

In order for genetic resources to be efficiently utilized in plant breeding programmes, it is first necessary to determine whether useful genetic variation exists in the material and secondly to develop the most cost-effective method of introducing the potentially useful genes into commercially acceptable material. Until recently, genetic analysis has relied on conventional segregation analysis in controlled crosses for qualitative traits determined by major genes, and on standard biometrical methods for quantitative traits controlled by polygenes. Now, however, the use of DNA techniques has opened up new possibilities for genetic analysis and breeding, and in this chapter the general principles and applications of these molecular marker-based procedures will be discussed.

In its specific sense, marker-assisted or marker-aided selection (MAS) is the use of appropriate easily recognizable genetic markers to facilitate or accelerate the selection of linked genes controlling useful agronomic traits. The nature of the marker loci themselves is not important and they are used simply to indicate the presence of useful alleles of genes of commercial or practical importance. However, the term has also come to include the wider use of markers to improve breeding programmes.

USES OF MARKER-AIDED SELECTION

In an ideal world, selection should be aimed directly at those genes which control the trait to be improved, but this assumes that genotypes at these target genes can be recognized easily, unambiguously and at an appropriate time without impeding the breeding programme. There are several reasons for using marker genes to improve selection efficiency.

Difficulties in Identifying Trait Genotype

The genotypes at a single-trait locus may be difficult to recognize because the desired alleles may have poor penetrance, be recessive, or interact with other genes or the environment. Environmental variation itself may impede accurate genotyping by causing the phenotype of different genotypes to overlap; this is particularly a problem with polygenic traits. Some phenotypes such as resistance to a particular pest, pathogen or abiotic factor, like drought or salinity, may only appear under particular conditions which are difficult to define or control. These difficulties may be more severe during the early stages of a breeding programme when plant numbers are too few to allow adequate replication or the breeder does not want to place his valuable yet scarce breeding material at hazard.

Earlier Selection

Another major reason why it may be valuable to use marker or indicator genes is to reduce the time from sowing to selection. Many traits of economic importance are only apparent in the mature plant and so may not occur for months or even years after sowing. This is particularly a problem with biennial species and long-lived crops such as trees. Indicator genes, on the other hand, may be detectable in seedlings within days of sowing, if not in the seeds themselves, so avoiding the waste of valuable resources involved in raising and harvesting plants, most of which may prove to be of no value to the breeding programme.

More Intense Selection

Selection at a juvenile stage, particularly among seedlings, may also allow much larger populations to be studied and hence more intense selection to be applied. Indeed, selection may be possible in tissue culture without raising plants or running trials.

Nondestructive Scoring

Many characters are scored before maturity and the act of scoring may prevent seed being collected. Imposition of selection for pests and diseases may result in plants which are less able to produce seed. The use of certain types of marker loci may only require the removal of small quantities of leaf or other material to permit full genotyping of the plant.

Linkage Drag

Conventionally plant breeders have used backcrossing with selection to introduce a useful allele, such as one conferring disease resistance, from one strain into another. The source of the allele is often an alien or exotic species, but may be a related cultivar. The F1 is backcrossed to the recurrent parent, and backcrossing is repeated for five to ten generations, the intention being to introduce the desired donor allele whilst returning to the recurrent parent's genotype at all other loci. At every backcross generation, the breeder selects for the phenotype of the allele to be introgressed whilst simultaneously selecting for the phenotype of the recurrent parent for all other traits. It was shown by Stam and Zeven (1981) that this procedure results in the leaving of very large amounts of donor genome associated with the selected alien allele. On average, about 20% of an average chromosome will consist of a donor segment even after ten backcross generations, while in any given line the actual size could be much more. Thus, as well as introducing useful alleles, the breeder will also be introducing very many undesirable alleles from the donor and so the gain in one trait may be at the expense of losses elsewhere. The use of marker loci can enable the breeder to reduce the size of this unwanted associated region as well as to accelerate the speed of return to the desired genotype of the recurrent parent.

Heterosis Potential

Heterosis, or hybrid vigour, is due mainly to the presence of dispersed dominant alleles in the parental cultivars. Breeders concerned with developing lines for hybrid production are often concerned to identify lines which differ at as many gene loci as possible. It is often assumed that this implies that the lines also have very different genotypes at marker loci and hence such loci can be used to assess the genetic distance between potential parents, but this can

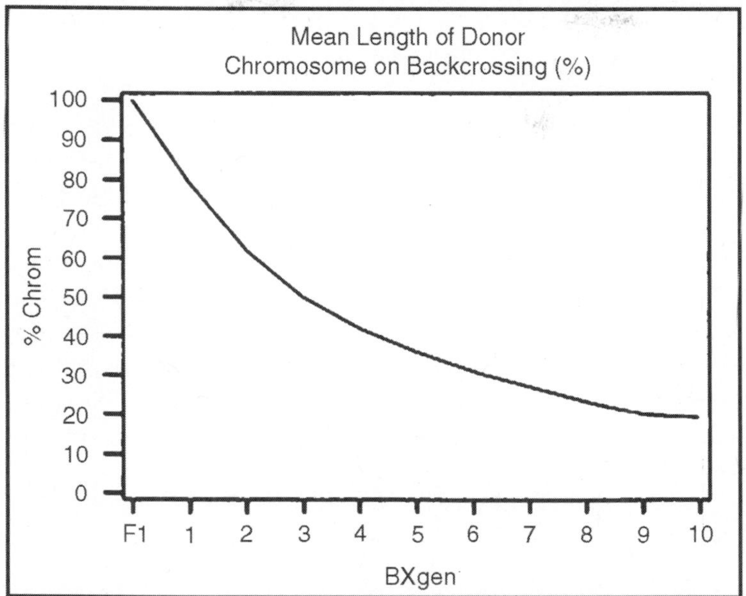

Fig. Linkage Drag.the Average Amount of Chromosome Surrounding a Selected GeneFollowing Backcrossing.Bxgen=Generation of Backcrossing from F_1

Quality Control

A major problem faced by most conservationists and breeders is to maintain the integrity of the material they are working with and to keep it free from contamination. Contamination can occur in various ways. It may arise through errors due to carelessness in labelling or sorting seed or plant material, mistakes in maintaining the pedigree, or as a result of outcrossing due to stray pollen from unrelated material. Such errors are difficult to recognize when only the gross phenotype is available as a guide, but genetic markers can provide a clear genetic fingerprint which can be used at any time to confirm the origin of the material. It can also be used to confirm that the intended cross or self has occurred, or to separate the desired progeny from a particular parent which might be a mixture of selfs and crosses.

Other Uses

Marker information can be used in other situations also: distinguishing haploids, diploids and aneuploids, for example, particularly following wide interspecific crosses or tissue culture. Doubled haploid lines produced by microspore or ovule culture can be recognized in this way and possible aneuploids arising during chromosome doubling can be identified and removed. Not only can genetic markers identify contaminant material but they also allow closely related or identical material to be identified in genebank collections, so reducing unnecessary duplication.

TYPES OF GENETIC MARKERS

Up to the mid-1960s the only easily usable genetic markers were those that produced clearly visible effects on the plant's phenotype such as colour, size, shape, disease resistance, etc.. All these were the result of mutations to alleles which had, obviously, a major effect on the development of the plant. Such mutants were rare and generally not desirable in commercial material although they may be selected by breeders of decorative plants for their novelty value. As genetic markers, however, they have many disadvantages. They often affect the fitness of the individual and could well have effects on important agronomic traits, i.e. they exhibit pleiotropy. Because they are rare, it would be necessary to bring several such mutants together in the same material in order to use them, and they would not normally be present in commercial cultivars or, indeed, in wild material. This not only takes time, but the combined effects of many such alleles would mitigate against their value as a genetic tool.

In the 1960s through to the 1980s, naturally occurring genetic polymorphisms at the protein level came to be easily recognized. Variation in enzymes and storage proteins was detectable on an electrophoretic gel for a very high proportion of gene loci which could be studied in this way and, unlike major mutants, such naturally occurring allelic variants had much smaller effects on fitness and hence were more common and existed naturally in populations. The number of such polymorphisms was, however, insufficient to provide the coverage of the genome desired by breeders and conservationists. The development of techniques during the late 1980s up to the present revealed the vast amounts of genetic polymorphism which exists at the DNA level and has revolutionized genetic analysis, opening up a whole new range of genetic tools.

As described elsewhere in this book, this variation arises through the existence of occasional base changes in the DNA which can be recognized by restriction enzymes or by primers used in analysis involving the polymerase chain reaction (PCR). Much of the DNA, possibly as much as 90% in many species, is noncoding and hence reliance on morphological or biochemical variants was only using the restricted variation that plants could tolerate within the coding parts of their DNA. Clearly, there are limits to the tolerance of such variation. Variation in the noncoding regions, whether in the intergenic regions or within introns, is probably under far less constraint by natural selection, and hence the number of polymorphic sites is potentially enormous. Because most of these sites are outside coding genes they are referred to as loci rather than genes; genes are coding regions whilst most marker loci are not. Depending on how the DNA polymorphism is studied, various types of marker are used. Restriction fragment length polymorphisms (RFLPs) have been the most widely used until now. They have the advantage of being codominant and hence all genotypes in a cross can be identified. On the other hand they require quite large quantities of DNA, and hence plant material, and generally rely on radioactive probes, although fluorescent techniques are becoming more popular.

As a result, RFLPs are expensive to use. The use of radioactive probes incurs safety considerations, as well as taking longer to visualize the bands. PCRbased markers are now being used more widely. Randomly amplified polymorphic DNAs (RAPDs), for example, require far less material because only small amounts of DNA are needed and the required sequences are then amplified. They do not involve radioactive probes and are both quick and cheap. On the other hand RAPDs are dominant although it is possible to distinguish heterozygotes from homozygotes by the amount of product. Amplified fragment length polymorphisms (AFLPs) offer even cheaper and more easily identified and extensive PCRbased polymorphism.

Because of the number of polymorphisms which exist on individual gels and the speed of production, computer software is necessary to scan the banding patterns on the gels and to analyse and assimilate the information that is generated. Other popular markers are microsatellites (short sequence repeats) and cleaved amplified polymorphisms (CAP)-based primers. All the genetic variants discussed above will be subsumed under the broad title of marker loci. They may be of little interest in their own right, and this is particularly true of the molecular polymorphisms; their value lies in their use as markers of more useful genes.

GENE MAPPING

An important, though by no means an essential, step in genetic analysis is to produce genetic maps of the marker loci. Such maps are often referred to as 'framework maps' because they provide a framework within which important genes can be located, as well as providing a means of comparing chromosome organization in other closely or distantly related species. There are two stages to mapping. Firstly, to arrange the markers in a linear sequence separated by an appropriate map distance, i.e. to construct a linkage map. The second is to relate the linkage maps to particular recognizable chromosomes. The latter is often the most difficult and generally not essential for breeding or conservation work. We will, therefore, concentrate on the former.

Chromosomes contain a single linear molecule of DNA and hence the markers on that chromosome occur at particular positions along that molecule. Typically, each chromosome contains 107 to 108 base pairs (bp), i.e. 104 to 105 kilo-base pairs (kbp) of DNA, while a typical structural gene, coding for a polypeptide chain, would be between 1 and 2 kbp long. Assuming only 10% of the genome is coding DNA, then a chromosome probably contains something of the order of 1000 to 10 000 genes. Fortunately, the fact that the chromosome is a linear molecule means that the genes need to be mapped in only one dimension.

Currently, the only useful method of gene mapping, at least as far as breeders and conservationists are concerned, relies on recombination between homologous chromosomes resulting from genetic exchange, which can be seen as chiasmata at diplotene through first metaphase of meiosis. A single chiasma on a particular chromosome results in half the gametes from such a meiosis being recombinant for that chromosome: half the gametes will contain a copy of that chromosome that is part maternal and part paternal in origin, while the remaining gametes will be entirely parental. Very few plant species have a sufficiently low number of clearly recognizable chromosomes that the number of chiasmata on a particular chromosome can be compared in different nuclei.

However, in many species the total number of chiasmata in each diplotene nucleus can be counted, at least in pollen meiosis, and the average number of chiasmata per nucleus calculated. The number and position of chiasmata on a particular pair of homologous chromosomes will vary from nucleus to nucleus; it is as though there are a large number of potential sites of exchange, but that in any given meiosis only a few sites are actually involved. The longer the chromosome the more potential sites there might be. There will invariably be one chiasma and there may be two, three, four or more; a typical set of results is shown in Table. A chromosome that has one chiasma on average is said to be 50 centimorgans (cM) long – a map length unit named after the American geneticist Thomas Hunt Morgan. This relates to the fact mentioned above that one chiasma results in 50% recombinant chromosomes. By extension, a chromosome with an average of 2.5 chiasmata is said to be 125 cM (i.e. 2.5 350 cM) long. It appears that, as a rough rule of thumb, each chromosome has on average two chiasmata and so is 100 cM long. It follows, therefore, that providing the haploid chromosome number is known (n), the total map length will be approximately 2 3 n 3 50 cM. The average number of chiasmata per nucleus has been calculated in many species and this rule generally holds.

Table. Chiasma Fraquency and Mapping.Chiasma Frequency in Chromosomes of Secale Cereale.

Number of chiasmata per chromosome	*Frequency*	*Percentage unrecombined Chromosomes*
1	0.267	50.0
2	0.716	25.0
≥3	0.017	12.5
Mean	–	22.0

This rough rule is important because it gives the geneticist an idea of the total genetic map length that should be expected as more genes or markers are mapped. It provides a guide to the extent to which the currently mapped markers cover the full genetic map. Because the map is based in chiasma units, it is not the same as a physical map. The distribution of chiasmata is not uniform over a given chromosome in a species and will vary with the chromosome and the species.

Although a knowledge of the number of chromosomes and, ideally, the mean chiasma frequency per nucleus for the species, indicates the total map length, the actual map has to be constructed from examining the frequencies of progeny in crosses; i.e. maps are constructed from information based on the consequences of chiasmata, namely recombination of genetic markers. Recombination of genetic markers is identified by the presence of gametes that contain recombined markers, and such gametes can be recognized from the phenotypes of progeny of a cross. The frequency of such gametes is an estimate of the recombination frequency between the two markers concerned. The closer two markers are to each other on the chromosome, the less likely a chiasma will occur between them and the less likely they are to recombine. Providing the two markers are sufficiently close that either none or one chiasma occurs between them but never more, then the recombination frequency as a percentage is equal to the map distance in centimorgans as described above.

The problem arises when the markers are sufficiently far apart for two or more chiasmata to occur between them in some meioses. It can be shown that not only does a single chiasma between a pair of markers result in 50% recombination between them, but so also do two, three or more chiasmata, on average. It is necessary to say 'on average', because there are a variety of possible consequences with two, three or more chiasmata in any given meiosis but, in practice, when one is looking at the progeny of a cross or self, each progeny will be the result of gametes from different meioses.

This fact implies that although map distance increases linearly with the number of chiasmata, the same is not true with the frequency of recombination, which can never be more than 50% for a pair of markers even though they might be 100 cM apart at opposite ends of a chromosome. These relationships are illustrated in Fig.. In order to overcome this problem, mapping functions have been devised to correct for multiple chiasmata. The most common of these are the Haldane (1919) and Kosambi (1943) mapping functions. Haldane's assumes that the probability of none, one, two or more chiasmata in a given interval follows a Poisson distribution, i.e. that the chiasmata are independent, random events. Kosambi's method allows for a certain degree of nonindependence in chiasma occurrences. It has long been known by cytogeneticists that chiasma interference occurs over quite large regions of a chromosome.Interference means that if a chiasma occurs at a particular point on the chromosome, the next one will never occur closer than some fixed distance, the interference distance, from it. Beyond this distance, there is a short distance in which the interference disappears and the next chiasma can then occur randomly outside this range. Observational data in various species of plants are confirming this and suggest that the interference distance is generally between 15 and 20 cM, i.e. 15 to 20% of a typical chromosome on either side of the chiasma. This has important and useful consequences for genetic and breeding work as will be shown later.

Recombination frequencies between pairs of markers can be scored in a wide variety of different crosses. The simplest to use are generations derived from an F1 because only two alleles are segregating, and their distribution in the chromosomes of the parents of the F1 can be determined. The generations that can be used are F2, backcrosses (Bc), recombinant inbred lines (RILs) or doubled haploid (DH) lines, with F2s being the most informative.

General formulae for calculating recombination frequencies are given by Allard (1956). In outbreeding species where F1s are not available, a given individual can be considered as an F1 and its selfed progeny as being an F2, even though the inbred parents do not exist. If it cannot be selfed but controlled crossing to another individual is possible, then again those genes segregating in the cross can be mapped although the situation is more complex. There could be from two to four alleles segregating at each locus and, with respect to any one locus, the cross could represent an F2, Bc1 or Bc2. Moreover, the distribution of alleles in the two parents cannot be known and has to be inferred from the progeny. This complexity is compounded by the large number of marker loci that may be segregating in any given cross.

Fortunately, a range of software packages are available to estimate these recombination frequencies, identify linkage groups, assign the markers to the most likely order and space them in map units (cM) on these linkage groups; examples are MAPMAKER and JOINMAP. Ideally the number of linkage groups should be equal to the chromosome number in the gametes but, unless the markers available provide a good coverage of the genome, it is likely that those markers on a given chromosome may appear as two or more separate linkage groups simply because the subsets are not sufficiently close for them to be recognized as being together on one chromosome.

A typical marker framework map is shown in Fig. for the chromosomes of *Brassica oleracea*. Clearly these maps provide a very detailed coverage of the chromosome. Many of the markers are very close, i.e. less than 10 cM, and over this distance it matters little which of the two mapping functions is used. With the more widely spaced markers, i.e. recombination frequency >15%, Haldane's function will exaggerate their distance apart because it will allow for more double crossovers than actually occur. As more markers are mapped, the total length should converge on the value predicted by the chiasma frequency, i.e. approximately 2 3 n 3 50 cM. It is important to remember that genetic map distances have standard errors which are dependent on the size and type of population used to construct the map, the mapping population. The map obtained is that which best fits, given the data and the assumptions underlying it, but there may well be other maps which also fit the data well, though are slightly less likely.

Any particular map always develops a greater aura of respectability once it is published, frequently without stating its reliability, and subsequent yet different maps are treated with suspicion. A recombination frequency of true value p has a standard error of ?[p(1 2 p)/N], where N is the size of the gamete population. For example, if p is 0.1 (i.e. RF = 10%) and$N$ = 100, the standard error is 0.03 or 3%. In other words the 95% confidence interval of the estimate lies approximately between 4% and 16%.

There are other reasons, apart from statistical sampling, why maps could be wrong. The most obvious is the accuracy with which the data are scored and recorded. Autoradiographs and banding patterns on gels can easily be misread and research workers are loath to discard data even though its interpretation is ambiguous. All data should be checked independently by two or more people and any ambiguities removed. Wrong scores will bias the data and often exaggerate map lengths as they suggest double recombination; wherever double recombination appears to have occurred in two short, adjacent intervals, the data should be checked. Changes in methylation rather than scoring errors could cause a restriction site to appear or disappear so giving the false impression of double crossing- over, or it could simply be an error.

Where maps appear to have major inconsistencies in different crosses, different chromosomal structural arrangements may be responsible, such as translocations or inversions. There is considerable evidence that chiasma frequencies, and hence recombination frequencies, may be quite different in male and female meioses even on the same plant although in other studies this may not be so.

Thus although the order of markers on a linkage group may remain the same, the relative distances between them may be quite different if the map derives from male versus female meioses. For example, in *Brassica* it appears that the genetic map obtained from a backcross where the F1 is the female parent is 60% longer than when the F1 is the male parent. It is also clear that environmental factors during meiosis, particularly temperature, can affect the number and distribution of chiasmata, and this could be very critical if the crosses are set up at different times of the year. Two or more different populations will almost certainly be segregating for different combinations of polymorphic markers, although some at least should be in common. These common ones can be used to overlay the two or more maps and a consensus map produced from an amalgamation of these. Again software is available to produce these consensus maps. Despite the various error-causing factors discussed above a considerable degree of consensus can be produced from such populations and, considerable similarity of chromosome sequence is conserved between even distantly related species, allowing the potential for cross-species genetic transfer.

It was stated above that the total map length should be dictated by the chiasma frequency, in so far as this can be accurately measured. When the first few genes were initially put onto genetic maps in the first 40 years of this century, they only covered a small part of the genome – except in wellstudied species such as *Drosophila* and maize. As more and more genes were placed on the map, so the total map lengths increased towards the asymptote dictated by the chiasma frequency. However, during the first decade following the use of molecular markers, the map lengths of some species appeared to exceed that expected, and it was even argued by some that the chiasma theory of recombination might be wrong.

No one would wish to argue that chiasma frequency data are free from error but there would have had to have been quite excessive underscoring to result in the disparity which was occasionally found. It would appear, however, that the excessive lengths were due in part to errors in scoring and to the use of small mapping populations, because subsequent map sizes have shown a progressive decrease towards the predicted asymptote. Scoring errors bias estimates of RF generally upwards, while RFs have to be large to be detectable in small populations. Our own experience with mapping in cereals and brassicas has also shown a progressive decrease in estimated map length as more markers can be found to fill some of the large gaps in the map and as errors are removed from the mapping data.

LOCATING GENES OF MAJOR EFFECT

The use of extensive molecular and other markers as described above provides a general framework map. Although this has value in its own right for comparing linkage groups in different species as well as possible structural variation within a species, the main value lies in providing a set of markers to locate genes of economic or special scientific interest.

Locating individual major genes is a straightforward development of the procedures above. Let us assume that a cultivar is identified which contains an allele of interest which changes the phenotype in a clearly recognizable fashion and which appears to segregate as a single gene in crosses to other cultivars. Using a knowledge of the positions of existing marker loci, a small subset of markers can be chosen which provide a reasonably even coverage of the genome, say every 20 to 30 cM, i.e. approximately five per chromosome. These then have to be shown to differ between the two cultivars and any which are monomorphic replaced by polymorphic markers. A rough position of the gene of interest can then be obtained by bulked segregant analysis. This involves taking the mapping population, e.g. a Bc, and dividing the plants into two groups depending on whether or not they show the phenotype of interest. Bulk samples of DNA are taken from each group, and the two samples assayed for the marker systems chosen as in the paragraph above.

Any marker which is unlinked to the gene to be located will produce similar banding patterns in both samples because it will be segregating independently of the target gene. Conversely, should one of the markers be very closely linked to the target gene, it will co-segregate and the two samples will show quite different patterns for that marker. In practice, complete co-segregation with any one marker is not very likely but, given a reasonable spread of markers over the genome, one or two are likely to show a strong association with the two groups. Thus although bands associated with both marker alleles will be found in each sample, their relative intensities will be quite different because of the association with the target gene.

Having located the chromosome and the region on the chromosome, the actual position can be identified in a segregating population using the principal marker associated in the bulked segregant analysis together with other polymorphic markers situated about 10 cM on either side of the principal marker. The target gene is then mapped with respect to the three markers. This approach can be used for a wide variety of populations and traits. Even if the trait has poor penetrance, the affected group will clearly show the marker banding pattern even if the nonaffected group consists of a mixture due to misclassification.

MAPPING GENES CONTROLLING QUANTITATIVE TRAITS

Quantitative traits such as yield, quality, height and flowering time, which are controlled by several genes and are greatly influenced by the environment, create particular difficulties for gene mapping. The problems arise because the genotype for the trait concerned can never be clearly identified from the phenotype; many different genotypes could produce the same or very similar phenotypes whilst the same genotype can result in different phenotypes depending on what may be elusive factors in the environment. Whereas with a single gene difference controlling a major effect, two or more Mendelian phenotypic classes might be recognized in ratios of 3:1, etc., quantitative traits resulting from the joint segregation of many genes show a continuous, often normally distributed range of phenotypes. The total phenotypic variation in an F2, for example, VP, is made up of genetic and environmental components, VG and VE. The genes underlying such traits are variously called polygenes, effective factors or, more recently, quantitative trait loci or QTL.

Segregation of particular marker loci with differences in the quantitative trait. Various mapping populations can be used as before, F_2, Bc, RIL, DH or open-pollinated full sib (FS) or half sib (HS) families, and the individuals or families are scored both for their phenotype for the quantitative trait(s) and for their genotype at the marker loci. The type of population that is used will depend on the traits studied and the breeding system of the plant species but, other things being equal, an F_2 population will normally provide the most information for a given size. Unfortunately, most traits of economic importance such as yield, quality and disease resistance are not usefully scored in an F_2 population because one is interested in how the material performs in something similar to the high-density monoculture that it will encounter in agricultural practice. Measuring yield in an F_2 population with spaced plants is straightforward and statistically very powerful but it may well not provide any insight into how yield is controlled in agricultural practice.

It is therefore necessary to use a plot structure for most traits in an attempt to simulate commercial conditions. For this reason most plant breeders will prefer to work with genotypes that can be extensively and easily replicated such as RILs, DHs or F_3s. It is also easier to explain the principles underlying QTL mapping using a population of DH lines, and so for both reasons the methodology will be illustrated with DH lines, although the same principles apply to all populations. Because several trait loci are concerned, probably on different chromosomes, bulk segregant analysis is of limited applicability, at least initially.

THEORY OF QTL MAPPING IN DOUBLED HAPLOID LINES

Doubled haploid (DH) lines can be produced parthenogenetically from an F1 by microspore or ovule culture and, in some species, such as wheat and barley, by wide outcrossing. Each original DH plant is derived from a single gamete, which subsequently becomes homozygous and disomic either naturally or by treatment with the drug colchicine, which inhibits spindle formation, and hence chromosome disjunction, at metaphase of mitosis. Selfed seeds from such a plant produce a DH line of identical individuals so the line becomes effectively immortal. If we consider an F1 that is heterozygous for a marker locus M1/M2 and for a linked QTL locus Q1/Q2, where the 1/2 indicates the allele with an increasing or decreasing effect on the trait, this will produce the gametes and hence the DH lines shown in Table.

If the DH lines are scored for some trait for which the Q+/Q+ homozygotes increase the trait above the overall mean, m, by a units while the Q2/Q2 homozygotes decrease the mean by a, then the means of the four gametic types are as shown in Table, where R is the recombination frequency between the QTL and the marker. From this it follows that the mean trait score of all those DH lines which are M1/M1 (M1M1) is m 1 a(1 2 2R), while the mean of all those DH lines which are M2/M2 (M2M2) is m 2 a(1 2 2R). In other words, half the difference between these two means (M1M1 2 M2M2) is a(1 2 2R).

Clearly, if the marker is so close to the QTL that it never recombines, *R* will be 0 and the marker difference will represent the QTL effect, *a*. Conversely if the two loci are unlinked, i.e. $R = 0.5$, then the difference will be zero. This effect is, of course, the basis of bulked segregant analysis. In general, the magnitude of the marker effect will decline linearly with the distance of the marker from the QTL in terms of recombination frequency, and so with several linked markers, their individual trait effects will be as shown in Fig.. Similar Arguments Apply to Other Populations Such as F_2s and RILs.

If the trait means of the markers and their map are known very accurately, and only one QTL existed on the chromosome, then as Fig. shows, it would be relatively easy to locate the QTL. Figure shows the distribution of flowering time among a number of doubled haploid lines of *Brassica oleracea* containing alternative alleles at a marker located very close to a QTL. In practice neither the map positions of the markers nor their means are known with very great accuracy because the number of DH lines and replicated plots are generally few and the heritability of quantitative traits is generally low. Moreover, linked QTLs create difficulties because their individual effects can combine either to create the impression of an intermediate, ghost QTL, or to conceal their effects altogether.

Various analytical procedures have been developed to maximize the efficiency and accuracy of locating individual QTLs and to separate linked QTLs. They all rely on the relationships described in the previous paragraph and use either maximum likelihood or weighted least squares regression procedures to estimate the QTL locations and effects that best fit the observed trait scores for each genotype. The main difficulty is that there are two unknown parameters with respect to each marker score; the QTL effect, *a*, and map position. This means that it is necessary, in all methods, to use an iterative approach which involves trying all possible QTL locations along each chromosome and identifying the position or positions of the QTL which best fit the observed data as indicated by the size of the likelihood or the residual regression variance. As a result, a very large number of statistical tests are performed with the concomitant risk of false positive results.

All methods provide estimates of QTL locations and effects with similar precision in terms of confidence intervals, and this precision drops rapidly as the heritability of individual QTLs declines, i.e. the estimates are accurate only when there are just two or three unlinked QTLs with large effects. For example, Hyne *et al*. (1995) showed that the 95% confidence interval for the location of a QTL contributing 10% to the phenotypic variance of an F2 could be as large as 35 cM. However, conservationists and breeders who are concerned with introducing QTL into commercial cultivars from other cultivars or more distantly related species, would normally only be interested in those cases where a few QTLs of major effect are to be manipulated. There is no evidence that dense maps are required for initial. QTL location. It is better to have large populations and a few (5–10) markers per chromosome. There is evidence that many of the larger QTLs located so far map closely to previously known major genes. This is often used as evidence that there are, in fact, very few QTLs, but this may well be misleading for several reasons. The QTL map locations are sufficiently imprecise that they could well appear spuriously close to candidate loci. Only QTLs of large effect will stand much chance of being located and, by definition, they are likely to be major genes. It is likely that many cases where QTLs have very large effects could be due to the chance association of alleles of like effect at several QTLs along a chromosome, which cannot be separated as individual effects.

Much effort has been devoted to increasing the precision of QTL location, in particular in improving the power of the test to reduce the failure to detect genuine QTL. A long-established problem with breeding for quantitative traits in plants is the genotype 3 environment interaction. Using QTL location techniques, it is now possible to explore these effects at the level of individual QTLs although it is often difficult to know whether genotype 3 environment or poor repeatability is responsible.

ALIEN GENE TRANSFER

Once useful genes have been located on a framework map, it is then possible to use the molecular markers to facilitate the transfer of the useful genes between species and strains in an efficient manner. As stated earlier, recombination is not a frequent event along a chromosome and, furthermore, two recombination events rarely occur within a distance of 15 to 20 cM of each other. It therefore follows that if a useful target gene (a major gene or a QTL) is known to be located within a region of chromosome flanked by two markers that are no more than 15 cM apart, then these markers can be used to follow and control the progression of the target gene through successive stages of a breeding programme. Clearly, the alleles at the marker loci have to differ in the donor and recipient cultivars, but any chromosome that contains A1 and B1 at the marker loci will also contain T1, the target allele to be transferred.

Should single recombination occur close to T, then the chromosome will contain A1 or B1 but not both. Providing the markers are less than 15 cM apart double recombination resulting in A1T2B1, and hence loss of T1, can be safely ignored. If the location of T is not accurately known, then three or more marker loci may be needed to be sure of safely bracketing the region containing T without fear of double recombination. There are conflicting requirements in marker-assisted gene transfer: the need to keep the bracketed region large enough to be sure of holding the gene to be transferred whilst not having it so large that too many other linked but undesirable alleles are transferred with it. Let us now consider the basic procedure of gene transfer. Traditionally, breeders have introduced a useful allele into their cultivars by backcrossing the original F1 to the commercial cultivar whilst selecting at each generation for the desired allele. As was shown earlier, this can result in a very large region of chromosome around the target gene surviving even after ten generations of backcrossing – a phenomenon known as linkage drag. With markers, however, it is possible to reduce linkage drag considerably whilst simultaneously reducing the number of backcrossing generations to two or three instead of the normal six to eight. Moreover, the amount of plant material raised at each generation can be reduced because much of the molecular genotyping can be achieved at the juvenile stage.

The aim is to hold the target allele, T1, heterozygous throughout the backcrossing process, by selecting for the flanking markers, whilst simultaneously selecting for the genotype of the recurrent, commercial cultivar at all other loci. The latter can readily be achieved by having as few as four or five well-spaced markers on all chromosomes and selecting for recurrent parent alleles at each generation.

Depending on the chromosome number, some individuals in the first backcross generation will be homozygous for several whole chromosomes whilst still being heterozygous for the target gene. Those which have most of the recurrent parent chromosomes are selected and backcrossed again, and if necessary the process is repeated for a third generation. For each generation, fewer markers need to be screened as those found to be homozygous in the previous backcross no longer have to be checked.

Unless the position of the target gene is very accurately defined, it is prudent to have several markers covering its likely position, again with the proviso that they need to be separated by no more than 15 cM. At the final stage of the introgression process, a backcross individual which is homozygous for most of the recurrent parent alleles is selfed or, if this is not possible, intercrossed with a similar genotype. From among the progeny, individuals are chosen which have different combinations of markers bracketing the target region as shown in figure.

These can be selfed again and homozygotes for the different sequences identified and multiplied for reassessment for the target trait. Some will fail to show the target trait because of recombination, but of those that do, the lines with the shortest sequence of donor markers are selected for further trials. The approach can obviously be adapted to the simultaneous introgression of several genes: no new principles are involved but the screening process becomes more elaborate. It is always possible at a later stage to reduce the length of the introgressed region around T by looking at other, more closely linked, markers. Similarly, it is wise to check the origins of the other chromosomes to avoid having introgressed unwanted regions through missing double recombinants in the central regions of individual chromosomes or single recombinants towards their ends.

If such errors are found they can easily be corrected by another round of backcrossing to the recurrent parent. When attempting to reduce the length of donor chromosome surrounding the target gene, it is important to realise that this should be done in two successive generations. If the sites of the required recombination events are less than 15 cM apart, then the simultaneous double event will not occur.

Even if they are far enough apart for the double event to occur, the probability of such an event may be too low to be practicable. If two recombination events each have a probability of 0.05 (i.e. 5%), their combined probability is at best 0.0025. Therefore, it would be necessary to genotype approximately 1200 individuals to be 95% sure of having at least one double recombinant. If it was tackled in two stages then it would require only the genotyping of 86 (= 28 in the first round plus 58 in the second; less being needed in the first round because the recombination could occur on either side of T), a very considerable saving in effort of 93%, or 1114 plants! Marker-aided selection has been successfully tried for quantitative traits in several crops while the efficiency of the procedure has been examined by Lande and Thompson (1990).

MAP-BASED GENE CLONING

Providing a target gene can be located to within a pair of marker loci, it should be possible to use standard cloning techniques to walk along the chromosome between the markers and to find the target gene. The feasibility of this depends on the distance between the markers and the ease with which the target gene can be recognized from among the clones.

Depending on the species, a genetic length of 1 cM can, on average, consist of as few as 280 kbp of DNA in rice, or 2220 kbp in maize. It is not always easy to locate a target gene with sufficient accuracy to be able to say that it is between two markers as close as 1 cM, and hence the actual lengths of DNA between them can often be much greater. With QTLs one can seldom achieve anything approaching this accuracy by existing methods of conventional mapping. Furthermore, these are average distances per centimorgan; some chromosomal regions could be much longer or, indeed, less.

Increases in mapping reliability can be obtained by using what are called near isogenic lines (NILs), which are lines obtained by selfing or backcrossing and are known to differ from some standard genotype by just a short defined section of chromosome. Providing this section is delineated by accurately mapped flanking markers, and the NILs which do and do not contain this region can clearly be shown to differ for the target gene, then the target gene has to be in that region. The genetic and physical proximity of the markers can be established by conventional mapping and by gene cloning. Map-based gene cloning involves starting from one of the markers and walking to the next by means of a series of partially overlapping cosmid clones. Cosmid clones are necessary because they can be up to 50 kb in length and hence require fewer cloning steps in order to cover the DNA between the markers (280 to 2200 kb per cM in the examples cited above). Identifying the target gene is more difficult. Clearly, if the gene is between the two flanking markers it has to be on one of the cloned regions and could possibly be on two overlapping clones.

It could be identified by transforming plants with each clone separately and identifying which transformant expresses the effect. If the gene concerned exhibits classical dominance it would be necessary to transform the homozygous recessive with the dominant allele or the transformant would not be recognized. Alternatively, the gene has to be introduced into a species that does not normally express the gene. Once the effective cloned fragment is identified, the actual location of the gene within the fragment could be identified by successively transforming sub-fractions of the clone or, following sequencing, by looking for potential open reading frames. In those cases where the target gene produces a known product, cDNA clones derived from tissue likely to be rich in the appropriate mRNA of the target gene can be used to identify likely sequences in the previous overlapping clones.

Once the gene has been identified by one of these methods, its structure and activity can be studied in detail. Disease resistance genes are obvious candidates for such studies as their action is very specific and their location on the genome of many species is well documented. However, unless suitable candidate loci.

3

Gene Transfer Methods in Plants

To achieve genetic transformation in plants, we need the construction of a vector (genetic vehicle) which transports the genes of interest, flanked by the necessary controlling sequences, *i.e.*, promoter and terminator, and deliver the genes into the host plant. The two kinds of gene transfer methods in plants are:

METHODS OF GENE TRANSFER IN PLANTS

To add a desired trait to a crop, a foreign gene (transgene) encoding the trait must be inserted into plant cells, along with a "cassette" of additional genetic material.

The cassette includes a DNA sequence called a "promoter," which determines where and when the foreign gene is expressed in the host, and a "marker gene" that allows breeders to determine which plants contain the inserted gene by screening or selection. For example, marker genes may render plants resistant to antibiotics that are not used medically (*e.g.*, agromycin, canamycin) or tolerant to certain herbicides.

Two methods are used to transfer foreign genes into plants. The first method involves the use of a plant pathogen called Agrobacterium tumefaciens, which causes crown gall disease in many species. This bacterium has a plasmid, or loop of non-chromosomal DNA, that contains tumor-inducing genes (T-DNA), along with additional genes that help the T-DNA integrate into the host genome. For genetic engineering purposes, Agrobacterium must first be "disarmed" so that it does not make the plant sick.

This is done by removing most of the T-DNA while leaving the left and right border sequences, which integrate a foreign gene into the genome of cultured plant cells. The second delivery method is a "gene gun," which fires gold particles carrying the foreign DNA into plant cells. Some of these particles pass through the plant cell wall and enter the cell nucleus, where the transgene integrates itself into the plant chromosome. Because both methods of gene transfer are fairly random, one must screen for the plant cells that contain the foreign gene

Vector-Mediated or Indirect Gene Transfer

Among the various vectors used in plant transformation, the Ti plasmid of Agrobacterium tumefaciens has been widely used. This bacteria is known as "natural genetic engineer" of plants because these bacteria have natural ability to transfer T-DNA of their plasmids into plant genome upon infection of cells at the wound site and cause an unorganized growth of a cell mass known as crown gall. Ti plasmids are used as gene vectors for delivering useful foreign genes into target plant cells and tissues. The foreign gene is cloned in the T-DNA region of Ti-plasmid in place of unwanted sequences.

To transform plants, leaf discs (in case of dicots) or embryogenic callus (in case of monocots) are collected and infected with Agrobacterium carrying recombinant disarmed Ti-plasmid vector. The infected tissue is then cultured (co-cultivation) on shoot regeneration medium for 2-3 days during which time the transfer of T-DNA along with foreign genes takes place. After this, the transformed tissues (leaf discs/calli) are transferred onto selection cum plant regeneration medium supplemented with usually lethal concentration of an antibiotic to selectively eliminate non-transformed tissues. After 3-5 weeks, the regenerated shoots are transferred to root-inducing medium, and after another 3-4 weeks, complete plants are transferred to soil following the hardening (acclimatization) of regenerated plants. The molecular techniques like PCR and southern hybridization are used to detect the presence of foreign genes in the transgenic plants.

Vectorless or Direct Gene Transfer

In the direct gene transfer methods, the foreign gene of interest is delivered into the host plant cell without the help of a vector. The methods used for direct gene transfer in plants are: Chemical mediated gene transfer, *e.g.*, chemicals like polyethylene glycol (PEG) and dextran sulphate induce DNA uptake into plant protoplasts. Calcium phosphate is also used to transfer DNA into cultured cells. Microinjection where the DNA is directly injected into plant protoplasts or cells (specifically into the nucleus or cytoplasm) using fine tipped (0.5 - 1.0 micrometer diameter) glass needle or micropipette. This method of gene transfer is used to introduce DNA into large cells such as oocytes, eggs, and the cells of early embryo.

Electroporation involves a pulse of high voltage applied to protoplasts/cells/tissues to make transient (temporary) pores in the plasma membrane which facilitates the uptake of foreign DNA. The cells are placed in a solution containing DNA and subjected to electrical shocks to cause holes in the membranes. The foreign DNA fragments enter through the holes into the cytoplasm and then to nucleus. Particle gun/Particle bombardment - In this method, the foreign DNA containing the genes to be transferred is coated onto the surface of minute gold or tungsten particles (1-3 micrometers) and bombarded onto the target tissue or cells using a particle gun (also called as gene gun/shot gun/microprojectile gun).The microprojectile bombardment method was initially named as biolistics by its inventor Sanford. Two types of plant tissue are commonly used for particle bombardment- Primary explants and the proliferating embryonic tissues.

Transformation - This method is used for introducing foreign DNA into bacterial cells, *e.g.*, E. Coli. The transformation frequency (the fraction of cell population that can be transferred) is very good in this method. *E.g.*, the uptake of plasmid DNA by E. coli is carried out in ice cold $CaCl_2$ (0-50C) followed by heat shock treatment at 37-450C for about 90 sec. The transformation efficiency refers to the number of transformants per microgram of added DNA. The $CaCl_2$ breaks the cell wall at certain regions and binds the DNA to the cell surface.

Conjuction - It is a natural microbial recombination process and is used as a method for gene transfer. In conjuction, two live bacteria come together and the single stranded DNA is transferred via cytoplasmic bridges from the donor bacteria to the recipient bacteria. Liposome mediated gene transfer or Lipofection - Liposomes are circular lipid molecules with an aqueous interior that can carry nucleic acids. Liposomes encapsulate the DNA fragments and then adher to the cell membranes and fuse with them to transfer DNA fragments. Thus, the DNA enters the cell and then to the nucleus. Lipofection is a very efficient technique used to transfer genes in bacterial, animal and plant cells.

Selection of Transformed Cells from Untransformed Cells

The selection of transformed plant cells from untransformed cells is an important step in the plant genetic engineering. For this, a marker gene (*e.g.*, for antibiotic resistance) is introduced into the plant along with the transgene followed by the selection of an appropriate selection medium (containing the antibiotic). The segregation and stability of the transgene integration and expression in the subsequent generations can be studied by genetic and molecular analyses.

USEFUL TRAITS BY GENETIC TRANSFORMATION AND NUTRITIONAL IMPROVEMENT

In addition to its basic calorific value, wheat with its high protein content is an important source of plant protein in the human diet. One of the prominent targets for the application of transgene technology for the nutritional improvement of wheat is targeted at enhancing the grain quality by, i) increasing the protein content, ii) increasing essential amino acids such as lysine, iii) increasing the high molecular weight (HMW) glutenins to improve breadmaking properties of wheat flour, iv) modifying starch composition and v) producing pharma-and neutraceuticals. Amongst the cereals, the flour of bread wheat,

Triticum aestivum, has a superior capability of forming leavened bread. This superiority stems from the structure and composition of its seed storage proteins, which upon hydration can interact to form gluten, an insoluble, but highly hydrated, visco-elastic aggregate that endows the wheat dough with its unique properties. Although the majority of wheat seed storage proteins participate in gluten formation, biochemical and genetic evidence has demonstrated that high molecular weight glutenin subunit (HMW-GS) plays a major role in determining the visco-elastic properties thereby determining bread making qualities.

The HMW glutenins are necessary to create a strong dough, which is essential for making high quality, yeast-raised breads. Strong dough traps tiny bubbles of carbon dioxide gas formed naturally by yeast during mixing and subsequent raising thereby enabling the dough to rise forming leavened breads. Dough strength and the ability to contain gas bubbles is known as visco-elasticity and is an important characteristic of wheat with respect to its end product quality. Increasing understanding of the molecular basis of dough visco-elasticity would thus help in developing strategies for minimizing the effect of unfavourable environmental factors on wheat end-use quality.

Genetic transformation of wheat is a key component in a scheme proposing a complete set of approaches to apply biotechnology to improve wheat quality via direct manipulation of HMW-glutenin genes. To alter the amount and composition of these proteins by genetic engineering, a gene encoding a novel, hybrid-subunit of HMW-GS under the control of native HMW-GS regulatory sequences was inserted into wheat using the biolistic approach (Blechl and Anderson, 1996). The HMW-GS 1Ax1 gene which is known to be associated with superior bread making quality, was introduced into the cultivar (Bobwhite) lacking this gene. The introduced 1Ax1 gene under the control of HMW-GS promoter was expressed at high levels and stability was maintained for several generations.

The results demonstrate the feasibility of manipulating the composition of wheat kernels by genetic engineering and have successfully made changes in both the levels and the types of seed storage proteins. This work also demonstrated the usefulness of tissue-specific promoters for the expression of transgenic proteins in the endosperm tissue of wheat. The prospect of achieving the elusive goal of nutritional improvement of wheat brightened further with the improvement in functional properties of wheat dough due to transformation of wheat with high molecular weight subunit genes. Transformation with one or two subunit genes results in stepwise increase in dough elasticity. The expression of a recombinant protein with x-and y-type HMW-GS subunit in transgenic wheat has resulted in altered gluten polymer assembly and composition. This study thus made it feasible to change the glutenin composition by expressing modified HMW-GS in transgenic plants.

The quality improvement studies were extended to durum wheat by He *et al.* 1999 with the introduction of one of the two subunits of HMW-GS subunit genes (1Ax1 or 1Dx5). Analysis of the expression of the additional subunits in the T2 generation using a mixograph, indicated an increase of dough strength and stability. This study demonstrated the feasibility of manipulating durum wheat for superior bread and pasta making qualities. Similarly, the overexpression of the HMW subunit 1Dx5, in transgenic wheat (T-aestivum) resulted in a four-fold increase in this proportion of component in the seed protein and also a corresponding increase in the proportions of the total HMW proteins and glutenins.

However, the overexpression of the 1Dx5 gene was found to be associated with a dramatic increase in dough strength by making it too strong, and, therefore, unsuitable for use in conventional breadmaking. Recently, Alvarez *et al.* (2000) introduced the HMW-GS genes 1Ax1 and 1Dx5 into a commercial cultivar of T-aestivum that already expresses five subunits. The overexpression of 1Dx5 gene increases the contribution of the HMW-GS to a level of 22% of the total protein content. In near future, molecular approaches including genetic transformation and marker-assisted selection will provide an opportunity for improving further the wheat processing qualities.

Modification of wheat starch is presently targeted in various laboratories to improve its potential utility. Wheat grain is predominantly composed of starch, which is a mixture of two polymers, the almost linear amylose molecules, and the heavily branched amylopectin molecules. The ratio of amylose to amylopectin in starch determines its physico-chemical characteristics and thereby its end-use. Wheat flour, low in amylose content is desirable for noodlemaking as it improves noodle texture. In wheat, scientists are thus working towards increasing the amylopectin content of starch thereby reducing amylose content leading to the formation of a value added, low amylose flour. The starch branching enzymes catalyse formation of 1, 6-linkages in the glucan polymer and control the amount of amylopectin produced. In recent years, efforts are underway towards detailed characterization of the starch branching enzymes and starch synthases. This increasing information and characterization of the various components of starch biosynthesis will enable researchers to make rational design of novel starches and alteration in starch levels in other crop plants as well.

Wheat is widely used as an animal feed also for non-ruminants in several developed countries of the world. The phytase of Aspergillus niger is used as a supplement in animal feeds to improve the digestibility and also to improve the bioavailabilty of phosphate and minerals. The phyA gene from Aspergillus niger, encoding for the phytase enzyme has been successfully expressed in transgenic wheat lines by the microprojectile bombardment of immature embryos. The constitutively overexpressed phytase was found to accumulate at high levels in the endosperm. This work may thus open newer avenues for a wider applicability of wheat. Wheat is also an ideal system for the production of novel compounds due to its excellent storage properties and the existence of an efficient processing industry. The potential products include the high value, low-volume compounds such as biologically active proteins and peptides, as well as, high volume, low-cost raw materials for packaging and building. The production of recombinant antibodies in rice and wheat was recently reported with the expression of a medically important, single chain Fv recombinant antibody against the carcino-embryonic antigen.. The recombinant antibody was targeted to the plant cell apoplast and endoplasmic reticulum and was detected in the leaves and seeds of wheat. More significantly, the recombinant antibodies remained active after prolonged seed storage at room temperature thus opening avenues for the further exploitation of wheat for producing high-value, novel compounds.

ENGINEERING NUCLEAR MALE STERILITY

The production of hybrids is an essential component of crop breeding programmes but till date, hybrid wheat has remained elusive! The development of a suitable hybridisation system for wheat requires a high degree of male sterility in all parts of the female parent to avoid self-fertilization. De Block and coworkers have developed a nuclear male sterile system in wheat by introducing the barnase gene under the control of a tapetum specific promoter, the expression of which prevents normal pollen development at specific stages of anther development. This system employed the ribonuclease-inhibitor barstar gene to restore the fertility of male sterile plants. To avoid complicated gene integration patterns, the target tissues were incubated on niacinamide containing medium before bombardment. The authors suggest that the enzyme poly (ADP-robose) polymerase (PARP), which plays a key role in the processes of cell division and recombination, is inhibited by niacinamide, thereby resulting in simple integration pattern of the transgene. Expressing the barnase gene at specific stages of anther development destroys the tapetum, thereby preventing normal pollen development and causes pollen sterility.

RESISTANCE TO BIOTIC STRESS

The integration of transgenic approaches with classical breeding techniques offers a potential chemical-free and environment-friendly solution for controlling pests and pathogens. Wheat is attacked by a number of viral, bacterial and fungal pathogens and also by insect and nematode pests. Introgression of genes from the wild relatives exhibiting resistance to pests and pathogens has been successfully utilized over the years for generation of resistant varieties in wheat. With the development of plant transformation techniques, newer avenues for creating disease resistant and insect resistant crops have been created. Coupled with this is the availability of novel transgenes encoding highly potential anti-microbial peptides, defense-related proteins and enzymes for the production of anti-microbial compounds in crop plants, have greatly enhancing the possibility of engineering crop plants for resistance to pests and pathogens. Fungal pathogens of wheat cause severe crop damages by infecting the spikes, leaves and roots. Amongst the various strategies to introduce fungal resistance by transgenic approach, strengthening the host plant defense by genetic manipulation hold tremendous potential. Biochemical and structural responses against fungal attack include reinforcement of plant cell wall, accumulation of phytoalexins with microbial toxicity, ribosome-inactivating proteins (RIP) that inhibit protein synthesis, antimicrobial peptides and synthesis of other PR proteins. Genetic engineering allows the expression of foreign genes from distant unrelated species as well as the modification of the usual pattern of expression of an already present gene.

Most of the works on genetic engineering of wheat for resistance against biotic stress have focussed on developing protection against fungal pathogens. Introduction genes encoding for chitinases from barley resulted in increased resistance against Erysiphe graminis. Bliffeld *et al.* (1999) reported the adverse effect of a ribosome inactivating protein on plant regeneration and development. Nonetheless, moderate protection against Erysiphe graminis was reported with the transformation of wheat with a gene encoding for a ribosome inactivating protein from barley. The genes encoding for thaumatin like protein (TLP) and stilbene synthase in transgenic wheat have been shown to improve resistance of T1, T2 progeny plants against the fungal pathogens. An increase in endogenous resistance against Tilletia tritici was achieved with the introduction of virally encoded antifungal protein.

For the engineering of resistance against different pests and pathogens in wheat, genes encoding for viral coat proteins, antifungal proteins, and proteinase inhibitors have been successfully introduced. Most of the introduced genes confer increased resistance to the corresponding pests and pathogens in the transgenic plants. Introduction of canditate genes (lectin, proteinase inhibitor) for insect resistance into wheat have resulted in growth inhibition of insects on transgenic seeds, thereby decreasing the fecundity of insect population.

RESISTANCE TO ABIOTIC STRESS

Traditional approaches at transferring resistance to crop plants are limited by the complexity of stress tolerance traits, as most of these are quantitatively linked traits (QTLs). Nonetheless, the direct introduction of a small number of genes by genetic engineering offers convenient alternative and a rapid approach for the improvement of stress tolerance. Although, present engineering strategies rely on the transfer of one or several genes that encode either biochemical pathways or endpoints of signalling pathways, these gene products provide some protection, either directly, or indirectly, against environmental stresses. Drought is a major abiotic factor that limits crop productivity, thereby causing enormous loss. The genes encoding the late embryogenesis proteins (LEA) which accumulate during seed desiccation, and in vegetative tissues when plants experience water deficiencies have recently emerged as attractive candidates for engineering of drought tolerance. Transgenic approach has been used for successfully introducing and overexpressing the barley HVA1 gene encoding for a late embryogenesis abundant (LEA) protein by Sivamani *et al.* (2000) into wheat by particle bombardment.

Most of the transgenic lines tested displayed improvement in important agronomic traits, including total dry mass and water use efficiency, shoot dry weight, root fresh and dry weights, when plants are grown under soil water deficit conditions. In general, this investigation showed that the transgenic lines expressing the HVA1 gene had improved growth characteristics including an enhanced biomass yield under water deficit conditions. The discovery of novel genes, determination of their expression pattern in response to abiotic stress and an improved understanding of their roles in stress adaptation (obtained by the use of functional genomics) will provide the basis of effective engineering strategies leading to greater stress tolerance (Cushman and Bohnert, 2000).

TRANSGENE SILENCING IN WHEAT

Success at developing improved wheat cultivars through genetic engineering depends on stable and predictable expression of the inserted gene. However, gene silencing is a common phenomenon in the production of transgenic plants and needs to be effectively controlled for the desired results. This is all the more important because gene silencing is an important phenomenon involved during natural plant defense. Gene silencing is a complicated phenomenon as it includes both transcriptional gene inactivation and post transcriptional gene inactivation. The complex and the large genome size of wheat is expected to be prone to silencing by introduced transgenes.

With the development of transformation methodologies for wheat, more information has started to accumulate regarding the inheritance and expression of the introduced transgenes. Information regarding the long term stability of transgenes is thus of immense significance for the use of genetic manipulation as a tool in wheat crop improvement and is necessary for the introduction and subsequent expression of desirable agronomic trait. DNA methylation plays a significant role in establishing and maintaining an inactive state of the gene by rendering the chromatin structure inaccessible to the transcription machinery. Methylation of DNA is expected to result in reduced gene expression. Muller *et al.* (1996) studied the variability of transgene expression in clonal cell lines of wheat and found a negative correlation between the degree of methylation and marker gene expression. PEG-mediated approach was used to transform protoplasts isolated from suspension cultures of T-aestivum. The integration and expression of the selectable marker gene nptII fused with different promoters, namely, alcohol dehydrogenase, shrunken from maize, and actin1 from rice, was confirmed by Southern analysis and enzyme activity test. Transgenic cell lines under selection were 'protoplasted' and clonal callus lines were cultivated from genetically identical single cells without selection pressure.

A reduction/loss of marker gene expression was observed due to a reduction in the nptII transcript level and was seen to be associated with hypermethylation of the integrated DNA. The silencing effect was reversed by a 4-week culture phase on media supplemented with demethylation agent, 5-azacytidine, thus confirming the role of methylation in transgene silencing Demeke *et al.* (1999) studied the inheritance and stability of an Act1D-uidA: nptII expression cassette in T4 and T5 transgenic plants. Based on the histochemical localization of GUS activity, this study demonstrated the lack of any cytoplasmic effect on the inheritance of the transgene. The transgenes maintained a multiple integration pattern similar to that observed in the T1 generation. The transgenic plants which produced low gus and nptII activity in seeds had an intact expression cassette. Southern blot analysis of genomic DNA performed with the methylation sensitive enzyme, HpaII, showed the transgene in GUS negative plants to be highly methylated relative to the transgene in GUS positive plants. Some of the studies on gene silencing also indicate that multiple integration pattern and copy number is also associated with DNA methylation.

Cannell and coworkers studied the inheritance of gus and bar marker genes over three generations. The integration, inheritance and expression of these marker genes in the population studied were stable and predictable with a few exceptions. The inheritance of integration patterns was stable, and transmission/inheritance of the transgenes followed Mendelian ratios in a majority of lines. From this study, the authors claim that the transformation procedure, transgene integration, and marker gene expression had little effect on the transmission of transgenes to the subsequent progenies. However, over the three generations studied, a uniform, 'progressive' transgene silencing, specific to the gus gene, was observed. This gradual loss of the gus gene expression was not accompanied by a reduction in bar expression. Since in this case, the gus and the bar genes are located on the same plasmid, observations indicate a post-transcriptional gene silencing.

The silencing of HMW glutenins was also observed in transgenic wheat expressing extra HMW subunits. Characterization of six independent events involving the transformation of wheat HMW-GS genes, 1Ax1 and 1Dx5, in a cultivar expressing five subunits by particle bombardment resulted in partial or complete gene silencing. Silencing of all the HMW glutenin subunits was observed in two different events of transgenic wheat expressing the 1Ax1 subunit transgene and overexpressing the 1Dx5 subunit. Control of gene silencing thus remains a challenge for the immediate future. Meyer, (1995) advocated that the problem of gene silencing in wheat can be minimized by optimising methods for simple integration patterns, use of promoters and gene sequences isolated from cereals, use of matrix associated regions (MARs) or scaffold attachment regions (SARs), which insulate transgenes from surrounding chromatin, might help in reducing the gene silencing problem.

TRANSPOSON TAGGING IN WHEAT

Transposon mutagenesis has been widely exploited in various organisms to isolate genes that encode unidentified products. The maize activator (Ac) and Dissociation (Ds) elements are the best-studied transposable elements in heterologous host plants. Initial studies have reported the introduction and activation of the Ac/Ds elements into cultured wheat cells by using a wheat dwarf virus by particle bombardment. Takumi and coworkers has developed a transposon tagging system in wheat by introducing the Ac transposase gene under the CaMV 35S promoter into cultured wheat embryos by particle bombardment.

For the development of a transposon tagging system, embryos isolated from a stable Ac line were bombarded with a plasmid containing the maize dissociator (Ds) element located between the rice Act1 and the gus genes. The transient expression of the gus gene was observed after the excision of Ds elements. Southern and northern analysis of the T0 and T1 plants has exhibited the stable expression and inheritance of the Ac transposase gene. These results have demonstrated the precise processing of the maize Ac transposase gene and the synthesis of the transposase protein in transgenic Ac lines. The Ac transposase gene causes the transactivation and excision of the Ds in the Ac transgenic lines transformed with the maize Ds elements. Thus in near future we can expect traits of commercial importance to be tagged for a wider utility in important crop plants.

MARKER ASSISTED SELECTION IN WHEAT BREEDING

Conventionally, plant breeding depends upon morphological/phenotypic markers for the identification of agronomic traits. With the development of methodologies for the analysis of plant gene structure and function, molecular markers have been utilized for identification of traits. Molecular markers act as DNA signposts to locate the gene(s) for a trait of interest on a plant chromosome, and are widely used to study the organization of plant genomes and for the construction of genetic linkage maps. Molecular markers are independent from environmental variables and can be scored at any stage in the life cycle of a plant. Over the last several years, there has thus been marked increase in the application of molecular markers in the breeding programmes of various crop plants. Molecular markers not only facilitate the development of new varieties by reducing the time required for the detection of specific traits in progeny plants, but also fasten the identification of resistance genes and their corresponding molecular markers, thus accelerating efficient breeding of resistance traits into wheat cultivars by marker assisted selection (MAS). The availability of back-cross derived near isogenic lines (NILs) have also facilitated the analysis of various lines by using different marker systems. The introduction of alien genetic variation into wheat is a valuable and proven technique for wheat improvement. Wild relatives of wheat provide an enormous resource of new genes for wheat improvement, particularly disease and stress tolerance. These genes can be of use if recombined into lines adapted to the conditions of a particular region.

Initial studies on the application of molecular markers in wheat relied on the hybridisation based restriction fragment length polymorphism (RFLP) system-RFLP maps provided a more direct method for selecting desirable genes via their linkage to easily detectable markers thereby expediting the movement of desirable genes among varieties. The factors that had been instrumental for the use of RFLP in wheat was the limited number of polymorphisms observed among wheat lines and more significantly the availability of aneuploid stocks for the determination of chromosomal location of genes. In wheat, RFLP's have been used to map seed storage protein loci, loci associated with flour colour, cultivar identification, vernalization (Vrn1) and frost resistance gene on chromosome 5A, intrachromosomal mapping of genes for dwarfing (Rht12) and vernalization (Vrn1), resistance to preharvest sprouting, quantitative trait loci (QTL's) controlling tissue culture response (tcr), nematode resistance, milling yield, resistance to chlorosis induction by Pyrenophora tritici-repentis.

RFLP markers are also useful in selection programmes for resistance against pests and pathogens, which is otherwise labour and time consuming, and to detect homozygous individuals and have been used for resistance to barley yellow dwarf virus, resistance to wheat spindle streak mosaic virus, resistance against powdery mildew, resistance against leaf rust, resistance against cereal cyst nematode. The use of RFLP analysis in wheat has, however, been of limited use in the intervarietal analysis due to low level of polymorphism and the high cost for screening in breeding situations.

With the development of polymerase chain reaction (PCR) methodologies, Random Amplified Polymorphic DNA (RAPD) emerged as a convenient and effective technique for tracing alien chromosome segments in translocation lines. RAPD markers provide a useful alternative to RFLP analysis for screening markers linked to a single trait within near isogenic lines and bulked segregants. He *et al.* (1992) reported the development of a DNA polymorphism detection method by combining RAPD with DGGE (denaturing gradient gel elecrophoresis) for pedegree analysis and fingerprinting of wheat cultivars. RAPD markers can be converted to more user-friendly Sequence Characterized Amplified Region (SCAR) markers, that display a less complex banding pattern.

SCAR markers linked to resistance genes against fungal pathogens have been characterized in combination with RAPD and RFLP. In recent years, RAPD and other PCR based markers like Sequence Characterized Amplified Regions (SCAR), Sequence Tagged Sites (STS) and Differential Display Reverse Transcriptase PCR (DDRT-PCR) are increasingly being used for identification of desirable traits in wheat and related genera. These markers have been used in particular for disease resistance against viral and fungal pathogens and also for insect and nematode pests and have the potential of pyramiding of resistance genes for effective breeding programmes.

PCR based markers have been extensively characterized for genes of resistance against common bunt, Tilletia tritici, powdery mildew, Erysiphe graminis, leaf rust, Puccinia recondita, resistance against Hessian fly, Mayetiola destructor and Russian wheat aphid, Diuraphis noxia.

Simple sequence repeats or microsatellites are more promising molecular markers for the identification and differentiation of genotypes within a species. The high level of polymorphism and easy handling has made microsatellites extremely useful for different applications in wheat breeding. Microsatellites have also been used to identify resistance genes like Pm6 from Triticum timopheevii and Yr15 from breadwheat. In near future, molecular markers can provide simultaneous and sequential selection of agronomically important genes in wheat breeding programmes allowing screening for several agronomically important traits at early stages and effectively replace time consuming bioassays in early generation screens.

GENETIC TRANSFORMATION TECHNIQUES

A number of genetic transformation techniques have been developed since the first report of a transgenic plant. With few exceptions, these methods rely on the ability to transfer DNA to a single cell or small group of cells, such as nondifferentiated leaf tissue or meristems. Then some form of tissue culture is required to regenerate whole, fertile transgenic plants for the process to be completed.

Therefore, the first requirement for plant transformation is a reliable plant regeneration system using explants/tissues which are amenable to the DNA delivery system. This requirement was, more than any other factor, the greatest impediment to genetic engineering of members of the cereal and grain legume families – the two most important crop groups. Regeneration in most of these species is difficult, particularly for species like soybean and chickpea.

At best, regeneration is extremely genotype-dependent for species such as rice, where the *japonica* rices are easier to manipulate *in vitro* than the *indica* rices, which are more important on a worldwide basis. However, international effort in most of the important cereal species has largely overcome the problem, although there are still genotypes of a number of species which remain recalcitrant to regeneration. The cereals and pulses (grain legumes) proved somewhat recalcitrant to genetic transformation until the late 1980s, when rice became the first cereal to become reliably transformed at high frequency, with soybean first successfully transformed via *Agrobacterium* in 1988, although repeatable transformation was only achieved later using a microprojectile approach.

There are three basic transformation systems being routinely used in most genetic engineering laboratories internationally.

These are:

- *Agrobacterium*-mediated transformation.
- DNA uptake via protoplasts.
- Microprojectile bombardment.

Agrobacterium-mediated Transformation

Agrobacterium tumefaciens, the causal agent of crown gall disease, transfers a portion of its own DNA (T-DNA) to the plant, where it is stably incorporated into the plant genome. The T-DNA is delimited by specific 25 bp repeat sequences, known as the border sequences. None of the T-DNA is actually required for transfer, and whatever DNA is bound by the borders will be transferred, hence the pathogenic DNA can be effectively replaced with genes of interest, allowing delivery of these desirable genes to cells from which whole plants can be regenerated.

The ability of *Agrobacterium* to transfer genes to plants is known as virulence, and is largely controlled by the *vir* genes on the Ti plasmid, and to a lesser extent, the chromosomal virulence (*chv*) genes. The *vir* genes are induced by the plant's wound response, in particular the production of phenolic compounds such as acetosyringone. The Vir proteins then facilitate the identification, copying, transfer and targeting to the host nucleus, and incorporation on a plant chromosome of the T-DNA. Generally the simplest manner by which to replace the T-DNA is to remove it, leaving a 'disarmed' Ti plasmid with the *vir* genes, then introducing a binary plasmid which contains the gene(s) of interest within the border sequences. Hence, with the induction of virulence, the Vir proteins will effect the transfer of the engineered T-DNA.

Most dicotyledonous species are hosts to *Agrobacterium* and as such can be genetically engineered in this manner. *Agrobacterium*-mediated transformation remains the method of choice for most species of the *Solanaceae* and *Brassicae*, and other top 20 most important plants such as grape, orange, cotton, sugarbeet and apple. While there are a number of reports of successful transformation of important grain legumes including soybean, chickpea and

Phaseolus bean, there is generally great genotypedependence of such methods. It has been shown repeatedly that plant genotype 3 *Agrobacterium* strain interactions can vary from complete absence of transformation to 100% transformation within many species, including soybean, pea and sugarbeet.

Most important monocotyledonous species were once regarded as being outside the host range of *Agrobacterium*. In the late 1980s it became evident that some monocot species, including asparagus, yam and onion, were transformable using *Agrobacterium*. Early reports of *Agrobacterium*-mediated transformation of rice, maize and sorghum remained largely unrepeatable in these and other labs, until the first report of a high-frequency, repeatable, and painstakingly substantiated system was published for three *japonica* rice cultivars Chan *et al*., (1993) first reported stable transformation of rice using *A. tumefaciens*.

Progeny analysis and Southern hybridization demonstrated that it was possible to generate transgenic rice via *Agrobacterium*, albeit at low frequency. Hiei *et al*. (1994) improved the efficiency greatly by co-cultivating callus originating from scutella of immature embryos with a super binary vector then selecting on hygromycin, and produced over 400 primary transgenics of three cultivars. Detailed analyses of 29 plants and their progeny revealed that stable transformation was possible, and in most cases marker genes segregated in the expected Mendelian fashion.

In addition, these authors sequenced T-DNA/plant DNA junctions and found that incorporation occurred at similar sites to those found in transgenic tobacco plants. The 'super binary' plasmid contained *virB* and *virG* genes from the hypervirulent pTiBo542 plasmid, and, although it enhanced transformation frequencies, was not an absolute requirement for transformation. Basic requirements for *Agrobacterium*-mediated transformation are host recognition, virulence gene induction, the ability to select and proliferate transgenic cells, and the ability to regenerate whole, fertile plants from these cells. Host recognition and virulence gene induction require the correct carbon source and pH, and the presence of virulence-inducing molecules. Many wounded plant tissues produce signal molecules such as acetosyringone, which set in train the events of T-DNA transfer and integration into the host nuclear genome.

Protoplasts

Direct gene transfer has been the method of choice for monocotyledons, predominantly because most of these species have proved recalcitrant to *Agrobacterium*-mediated transformation. Protoplasts are cells with the cell wall enzymatically removed. In the absence of a cell wall, macromolecules such as proteins and nucleic acids may be taken up by such cells. The efficiency of uptake of DNA is greatly enhanced by the application of physiological shock, such as an ionic, osmotic or electrical current, which causes reversible permeabilization of the cell membrane. DNA has been shown to transfer to cells of most major cereals using the electric shock, or electroporation, methodology. If the DNA enters the nucleus and is stably incorporated into the plant genome, the cell is transgenic. This has been achieved with all major cereals (e.g. sorghum). One of the advantages of this methodology is that a large number of theoretically competent cells can be exposed to DNA at one time.

Regeneration of whole, fertile plants from cereal protoplasts has remained the greatest hurdle to electroporation-mediated transformation. Even when protoplasts are regenerable and DNA transfer is effected, it is difficult to combine the two. The physiological shock necessary for DNA uptake may often be enough to prevent subsequent cell wall synthesis or substantially reduce regenerability. As protoplast cultures usually require greater time *in vitro*, there may also be an increased frequency of somaclonal variation. Success has been achieved with regeneration of rice, wheat and maize but protoplasts remain a widely used method for rice only, with microprojectile-mediated transformation overtaking it in efficiency.

Microprojectile Bombardment

The potential for genetic transformation with microprojectiles was first reported by Klein *et al*. (1987), who achieved strong transient expression of a reporter gene in intact onion tissues. The first report of fertile transgenic plants and their progeny described stably transformed tobacco plants with Mendelian inheritance among selfed progenies. However, much of the focus with microprojectile bombardment has concentrated on cereals, due to the apparent inability of *Agrobacterium* to transfer genes to most cereal species, and the recalcitrance and genotype-specificity of cereals (with the exception of rice) to regeneration of fertile plants from transformed protoplasts.

The microprojectile bombardment, or 'Biolistic' (commercial description used by Bio Rad Labs), approach is a simple concept, based on the introduction of DNA-coated small inert projectiles into plant cells, not unlike the action of a microscale shotgun. The system is a modification of one used by virologists, whereby high-velocity microprojectiles are used to facilitate viral infection. To adapt this for transformation, the major alterations required were to reduce microprojectile size to reduce the extent of cell damage, to be able to perform the procedure under sterile conditions to exclude contamination by bacteria or fungi, and most importantly, to ensure that the target tissue be regenerable to produce whole, fertile plants.

Here lies one of the greatest advantages of the microprojectile approach over protoplast and *Agrobacterium*-mediated techniques. Effectively, any target tissue may be used for bombardment, so the most regenerable tissues can be used. In addition, there does not appear to be any genotype specificity in the ability to deliver DNA, which is demonstrably the case with *Agrobacterium*mediated transformation (e.g. Owens and Smigocki, 1988). The genotype dependence of plant regeneration remains the greatest hurdle in most species, although this is by no means the barrier that protoplast regeneration presents.

The microprojectiles, which are commonly an inert metal such as tungsten or gold, range in diameter from 1 to 4 mm; a velocity of approximately 250 m s-1 is required to penetrate cell walls and membranes. For successful stable transformation, the DNA must be delivered intact to the nucleus where the DNA will then be incorporated onto a chromosome. This requires that the velocity of microprojectiles be sufficient to penetrate cells, but not cause excessive damage.

For most of the cereals, the target tissue of choice is the immature zygotic embryo, particularly the scutellum, or embryogenic callus cultures, usually derived from zygotic embryos. Other targets such as embryogenic suspension cultures have also been used. As can be seen, the most important cereal species have been recently transformed using this approach. In addition, transgenic oats and rye have been produced with microprojectiles. Early microprojectile apparatus was based on a gunpowder charge for particle acceleration. Improvements in safety and reliability were made by using electrical discharge or helium gas to provide acceleration, and most laboratories currently use helium-based guns, such as the commercially available Biolistic gun.

Tissue sterility is maintained by performing all operations in a laminar flow cabinet with bombardment actually taking place in a sealed chamber under a slight vacuum. One of the drawbacks of the microprojectile approach is the cost of the Biolistic gun, hence other laboratories have developed simpler apparatus, such as the particle inflow gun (PIG), which relies on a solenoid to regulate helium gas inflow. The great advantage of the PIG is that it can be built for around US$1000. There are potential questions about the use of such apparatus and whether there is any contravention of international patents.

Alternative Transformation Methods

While the great majority of transgenic plants worldwide are being produced via *Agrobacterium*, microprojectile bombardment and, to a lesser extent, protoplasts, a number of alternative transformation methods have been trialled. Although not intended to be an inclusive list, these include:

- Virus-mediated transformation.
- Pollen pathway or intact inflorescence methods.
- Microinjection.
- Silicon carbide fibres.
- Electroporation of intact tissues.
- Electrophoresis of intact tissues.

While it is not possible to discuss the success, advantages and disadvantages of these methods, the subject has been recently reviewed by Songstad *et al.* (1995).

The pollen pathway has long been recognized as a potentially useful means of introducing genes to the germline, with its greatest advantage being the avoidance of tissue culture and the need to regenerate undifferentiated tissues such as callus or protoplasts. The absence of any time in plant tissue culture should also totally avoid the problem of somaclonal variation. There have been many difficulties with this approach, however, and there are no substantiated instances of genetic transformation via the pollen pathway.

More promising methods are the use of silicon carbide fibres and electroporation/ electrophoresis of intact tissues. Stable tissue transformation of tobacco and corn (maize) suspension cultures has been achieved following agitation with silicon carbide fibres. Stable cereal transformants have been produced by electroporating germinating rice seeds and immature embryos of corn, and by electrophoresis of germinating barley seeds. However, until some of these alternative systems can be made more reliable and efficient, it would appear that the two most widely used transformation systems will remain *Agrobacterium* and, in the case of cereals, microprojectile bombardment.

GENE IDENTIFICATION, ISOLATION AND TRANSFER

The conservation of plant genetic resources has long been espoused as an essential part of plant genetic improvement. Without genetic diversity it would not have been possible to make the genetic advances seen in crops worldwide this century. Indeed the very basis of the 'Green Revolution' lay in the ability to harness the benefits of plant breeding and crop physiology, and germplasm collections of important species such as wheat, rice and maize underpinned these improvements.

However, with the Green Revolution came a growing awareness that the use of high-yielding, genetically uniform cultivars across the world led to increasing genetic vulnerability. Much of the genetic gain, in terms of yield and/or adaptation to biotic and abiotic stresses, has been realized in the major crop species, most notably the cereals, grain legumes, oilseeds and, to a lesser extent, fibres and horticultural crops. The quality of many crops has altered such that for a number of species, the produce seen on modern Western supermarket shelves bears little resemblance to the commodities of the late nineteenth century. Some would say this has been to the detriment of crops such as tomatoes, the most commonly complainedabout fresh produce.

The source of most of this genetic diversity has generally come from within the cultivated species. Nevertheless there are many notable instances where extremely useful traits have originated from wild relatives, both at the interspecific and intergeneric levels. This is exemplified by crops such as wheat, with introgression from within and beyond the genus, most notably from rye, with the 1B/1R translocation furnishing resistance genes to fungal diseases such as powdery mildew, and stem and leaf rust. This translocation is widely used in many European wheat cultivars. Similarly, stripe rust resistance from *Agropyron elongatum* was transferred to wheat with a radiation-induced translocation and has been used extensively in Australian wheat cultivars.

Many of the disease resistance genes in tomato originate from the wild species of South America including *Lycopersicon hirsutum* (early blight resistance), *L. pimpinellifolium* (*Fusarium* wilt resistance) and *L. peruvianum* (TMV resistance) and introgressed into cultivated tomato. Many cultivars of banana and plantain are naturally occurring triploid and tetraploid interspecific hybrids of the Southern Asian and Southeast Asian species, *Musa acuminata* 3 *M. balbisiana*. There are edible diploid forms of *M. acuminata* (A genome) but not of *M. balbisiana* (B genome). All sugar-cane cultivars are complex aneuploid interspecific hybrids, and although based on the noble cane, *Saccharum officinarum*, cultivars contain genetic material from *S. spontaneum* and *S. robusta.*

A number of very important crops are allopolyploids of existing diploids. In some cases, such as oilseed rape or canola (*Brassica napus*), the alloploid is a naturally occurring hybrid of two cultivated species, namely cabbage (*B. oleracea*) and turnip (*B. campestris*), which arose from spontaneous chromosome doubling or the fusion of unreduced gametes. More, however, are the result of some historical hybridization involving species which are of little or no commercial importance. Cotton (*Gossypium hirsutum*), for example, is an allotetraploid derived from the diploid progenitors *G. herbaceum* and *G. thurberi*. As a result, the diploid progenitors are potentially useful sources of genetic variation, as long as sexual introgression of wild genes from this secondary gene pool is possible.

One particularly illustrative example of the importance of germplasm conservation is the alpine weed species, *Arabidopsis thaliana*. Although of no intrinsic economic value, this species has played a central role in the advances made in plant molecular genetics in the past decade. The species has great attractions as a model plant species: it has a very small genome, with little repeat sequence DNA; it is rapid cycling, with only 6–8 weeks generation turnover time; it takes up little space, even for highly replicated experiments; and it can be crossed/selfed and therefore is amenable to many experimental designs.

In 1983, the first report of a genetically engineered plant was published. A transgenic tobacco plant had been produced by utilizing genetically manipulated strains of *Agrobacterium tumefaciens*, introducing a bacterial gene encoding neomycin phosphotransferase, which rendered the plant resistant to a group of antibiotics, including kanamycin. Shortly after, similar plants were produced using a direct gene transfer system based on protoplast DNA uptake. In the decade or so since, a plethora of different methods and modifications of methods have been applied to produce transgenic plants of most important crop species.

With the development of plant genetic engineering technology, it is necessary to reassess the basis for plant genetic resource conservation. No longer is the gene pool for a species such as tomato within the *Lycopersicon* genus or even the *Solanaceae*. Genes may now be transferred to tomato not only across the family barrier to include all plant species, but from sources outside the plant kingdom including bacteria, fungi, viruses and animals.

Indeed transgenic tomatoes have been produced with sequences from viruses, bacteria (herbicide resistance to phosphinothricin from *Salmonella typhimurium*), antisense sequences of existing tomato genes, specifically designed to downregulate the expression of an endogenous gene (downregulation of the fruitripening enzyme polygalacturonase) and fish (improved cold tolerance with a gene encoding antifreeze proteins). Hence, the tertiary gene pool for any transformable plant species can now be regarded as taking in all life forms.

TARGET TRAITS

Conventional plant breeders have, depending on species of interest, targeted a wide range of traits for improvement, including yield, resistance or tolerance to biotic and abiotic stresses, adaptation to edaphic and climatic conditions, and product quality. With the possible exception of yield, there are examples of all other traits which have been manipulated using genetic engineering techniques. Hence it can be generally assumed that virtually all traits which are being manipulated via conventional breeding may be altered using molecular methods. However, with the wider gene pool accessible with genetic engineering, other traits not able to be manipulated via conventional means can be considered.

GENETIC ENGINEERING TECHNOLOGY

Genetic manipulation for biotechnological plant improvement relies on the ability to modify genes as DNA sequences. This requires utilization of standard, well-documented techniques. Enzymes are the basic tool kit of the genetic engineer, and can be used for basic functions of manipulating nucleic acids in such a way as to copy (DNA polymerase), transcribe (RNA polymerase), cut at specific sites (restriction endonucleases), cut and degrade in a nonspecific nature (nucleases) and join pieces together (DNA ligase). DNA libraries are another important resource for genetic manipulation as they represent a means of storing and screening sequences from an entire organism. The two basic types of library are the cDNA library and the genomic library. A cDNA (complementary DNA) library is based on expressed sequences. Such a library is made by extracting mRNA from a particular plant tissue of interest, then making first single-stranded DNA (via reverse transcriptase) then double-stranded DNA (via DNA polymerase). A cDNA library will then represent only those genes which are expressed in the tissue of interest (or in response to a stress of some sort), and will be devoid of introns and regulatory sequences.

A genomic library represents the entire genome (often including the organellar genomes), and will include all sequences whether they are expressed in the donor tissue or not, and even sequences which are not expressed at all. Genomic libraries will therefore include highly repeated sequences such as rRNA genes and nonexpressed simple sequence repeats (SSRs), as well as genes and their regulatory sequences and introns. Genomic libraries may be cloned and stored in lambda bacteriophage, cosmids or larger fragments which will often include multiple genes in bacterial artificial chromosomes (BACs) and yeast artificial chromosomes (YACs). BAC library inserts may average 150 kb, whereas YAC inserts are larger on average, at up to 1 Mb of DNA.

Overall, the ability to genetically engineer a plant relies on:

- A genetic transformation system for transferring DNA into plant cells.
- A plant regeneration system to produce plants from the transgenic cells.
- A gene construct which can modify the plant phenotype in a desirable and controlled manner.

CLONING OF PLANT TRAITS

Plant gene cloning is currently limited to simply inherited single gene traits. The transfer of genes to plants is limited to single genes or, at most, two to three genes encoding a particular, well-characterized biochemical pathway. Manipulation of multigenic traits is currently only feasible within the existing gene pool, sometimes made more efficient by markerassisted selection.

When traits are regarded in the molecular genetic sense, often the coding sequence is the major consideration. Obviously, the coding sequence is of major importance as this is the template for determining the amino acid sequence of the final protein or enzyme which determines the phenotype. However, a major factor in controlling the phenotype is the presence of regulatory sequences surrounding the coding sequence in the form of upstream (promoters and enhancers) and downstream (terminator) regulatory sequences and, in many eukaryotic genes, introns.

When traits are being manipulated via genetic engineering, the role of these regulatory sequences cannot be trivialized. Put simply, these sequences control such things as the level of gene expression, tissue/organ specificity, developmental expression and response to particular abiotic/biotic stimuli. The regulatory sequences that have been subjected to greatest attention in plants are the upstream promoters. For proper expression of a seed storage protein, for example, the gene of interest must be under the control of a promoter which expresses in developing endosperm (monocots) or cotyledons (dicots). Expression of other genes may be required only in anthers, root hairs or photosynthetic tissues, or in response to a wound such as that caused by a chewing insect or penetration of a fungal infection structure.

In the earliest cases of transgenic plants, introduced genes were placed under the control of constitutive promoters of bacterial (octopine or nopaline synthetase) or viral (cauliflower mosaic virus (CaMV) 35S) origin. These promoters have been found to give adequate expression of selectable marker and reporter genes in dicotyledonous species. However, expression is commonly not of a sufficient level in monocots, especially cereals, for good selection with antibiotic or herbicide resistance genes. And, as already discussed, such a constitutive pattern of gene expression is commonly not desirable for most traits. For cereals, a number of other promoters have been developed to achieve high-level constitutive expression of selectable markers or reporter genes. These include sequences cloned directly from cereal tissues such as the ubiquitin and rice actin 1 promoters and synthetic promoters involving various spliced sequences from a range of sources such as the Emu promoter.

However, to properly control gene expression such that the introduced character is most efficiently targeted without other pleiotropic effects, it is generally found that regulatory sequences of plant origin are desirable. Further to this, expression is usually best controlled when a cereal promoter is used in a cereal or a dicot promoter in a dicot. The most widely used means of cloning such regulatory sequences is by using cDNA libraries from specific tissue types and looking for abundant signals. Naturally, cDNA libraries will not contain the regulatory sequences, hence the abundant cDNA clones must then be used to screen a genomic library so that the regulatory sequences can be identified and cloned.

METHODS FOR CLONING PLANT TRAITS

Cloning Based on Knowledge of Protein Structure/sequence

Many of the first plant traits to be cloned were those in which some degree of biochemical understanding preceded any work with DNA. The understanding may have come from plant systems or from outside the plant kingdom. Most of the cloned genes in these cases were enzymes which were part of a well-characterized biochemical pathway or had a protein structure which was otherwise well known, such as seed storage proteins. Cloning genes based on a knowledge of the protein sequence can be achieved by working backwards from the gene product (amino acid sequence) to the coding sequence or the

DNA template. Such an approach has been used for cloning many seed storage proteins, such as the 2S sulphur- rich proteins from Brazil nut.

Genes of plant origin have been used in transgenic plants to improve resistance to insect pests. One of the first to be characterized was the cowpea trypsin inhibitor, a protease inhibitor which acts as an antifeedant to a range of insect pests. The protein is naturally produced in seeds of some genotypes of cowpea (*Vigna unguiculata*), which was shown to confer lepidopteran insect resistance when expressed constitutively in tobacco. Another such report from the same group at Durham

University described the use of a snowdrop lectin gene to confer aphid resistance in transgenic tobacco, the first report of genetically engineered resistance to a sucking insect in plants. The lectin from snowdrop (*Galanthus nivalis*), known as GNA, has been demonstrated to be effective against two major rice pests, the brown planthopper and the green leafhopper, as well as the peach potato aphid. When expressed at 0.1% of total leaf protein in tobacco, GNA is antimetabolic to homopteran pests. These levels of protein were accumulated in leaf tissues when placed under the control of the CaMV 35S promoter. However, Hilder *et al.* (1995) postulate that better control would be achieved with the use of a phloem-specific promoter, such as the sucrose synthase 1 promoter from rice. Another significant group of genes which have been cloned based on prior knowledge of the end product are the seed storage genes, including both seed storage proteins and lipids. In both cases there was considerable biochemical knowledge of the pathways involved in biosynthesis and storage. For seed storage proteins, the genes can be cloned based on the knowledge of the amino acid sequence of the mature protein and targeting signal or transit peptide. Molecular genetic manipulation of vegetable oils is a major area of research in the developed world. While approximately 90% of vegetable oils are used for human consumption, there are emerging markets for industrial applications including lubricants, paints, cosmetics, plasticizers and soaps.

With increasing emphasis on quality and dietary considerations, there is a desire in developed countries to substantially alter the types and balances of fatty acids found in the major oil crops. Six crop species (soybean, oil palm, rapeseed/canola, sunflower, cottonseed and peanut) account for 84% of the world's vegetable oil production, with the major fatty acids produced being palmitic, stearic, linoleic, linolenic and oleic acids. However, almost 200 different fatty acids are produced by plants, mostly occurring in nondomesticated plant species. Hence, genetic manipulation of the major crop species, either via conventional breeding or genetic engineering, has the potential to modify the fatty acid types produced and stored by seeds such as soybean and canola.

The biochemistry of fatty acid biosynthesis is well established for many of the major pathways, with enzymes identified which control the chain length of fatty acids, and the degree of saturation, or number of double bonds in the fatty acid chain. The oils stored in seeds have been altered by downregulation of particular enzymes using antisense RNA techniques. The oleic acid content of rapeseed oil has been increased to 83% (from 62% in conventionally bred cultivars) by downregulating the enzyme D12-oleate desaturase using antisense RNA. A similar result has been achieved using co-suppression, which has produced rapeseed oil with 87% oleic acid content. Genetic engineering approaches have also been used to introduce totally novel fatty acids to major oilseed crops. Petroselinic acid, an isomer of oleic acid, is found in abundance in the spice plant, coriander (*Coriandrum sativum*). This oil has uses in cosmetics and pharmaceuticals, and its oxidation leads to the formation of lauric acid, used in detergents, and adipic acid, which is used in nylon production.

Hence this is a potentially valuable oil which may enable soybean and canola crops – i.e. renewable resources – to replace some of the nonrenewable carbon fuels which are currently used for many industrial polymer and lubrication applications. A cDNA clone from coriander encoding an acyl-ACP desaturase has been genetically engineered into tobacco, resulting in the accumulation of petroselinic acid. Oil crops such as rapeseed and soybean could be genetically modified to produce this fatty acid on a commercial scale. Hence, here is an example whereby a minor crop plant has become a donor of genetic diversity which will enable the commercial production of a very useful array of products for food and industrial purposes.

Cloning Genes Known Only by Phenotype

However, a more challenging task has been to move beyond proteins and pathways about which sufficient biochemistry is known into cloning genes known by phenotype only. By far the majority of genes fit into this category, which includes many of the most important plant traits such as disease resistance, adaptation to abiotic stresses, phenology and flowering behaviour and fertility.Only in the past few years have genes for traits known only by phenotype been cloned. One of the most interesting areas of plant biology is the understanding of plant–pathogen interactions at the molecular level. The major methods used for cloning genes known only by phenotype are:

- Transposon tagging.
- Map-based or positional cloning.
- T-DNA tagging.

There have been a number of plant genes cloned using all these methods or modifications of them. There are also examples where a combination of methods has been used to expedite the rate of identification of the unknown sequence.

Transposon Tagging

The proposal that specific pieces of DNA are able to autonomously excise from a particular locus and insert, theoretically at random, into another linked or nonlinked locus was first proposed by Barbara McClintock (1949, 1950). Transposable elements are now accepted as ubiquitous, and are particularly well characterized in organisms such as *Escherichia coli*, *Drosophila* and maize.

In two plant species, maize and the snapdragon (*Antirrhinum majus*), transposable elements are sufficiently well characterized to be used for mutagenesis and gene tagging. The best understood system is the maize *Ac/Ds* (*Activator/Dissociation*) family of elements. The behaviour of the *Ac* element is such that it can excise autonomously from its locus, under the control of its own transposase, and insert into another chromosomal position. Where the *Ac* element is in or near a gene, that gene is inactivated (i.e. not expressed), and will appear phenotypically as a recessive mutation. When *Ac* transposes (excises and reinserts at another locus), the gene is commonly returned to normal function, and hence is termed a revertant.

Should this *Ac* element then insert into another gene (transcribed or regulatory sequence), then a new recessive mutation will manifest. Transposition can occur either germinally, which results in an altered phenotype for the whole plant, or somatically, which results in a mosaic mutation, and can often be detected as a sectoring of plant or seed colour. The *Ds* element is actually a group of elements derived from defective *Ac* elements, with a common feature being their lack of effective *Ac* transposase.

Table. Plant Transposable Elements which have been used for Transposon-Taggine

Element	*Species*	*Type of Element*	*Size*
Ac	Maize	Autonomous	4.6
Ds	Maize	Nonautonomous	0.5–30]
Tam1	Snapdragon	Autonomous	15
Tam2	Snapdragon	Nonautonomous	5
En/Spm	Maize	Autonomous	8.3

Hence these elements are nonautonomous, and will only transpose in the presence of an *Ac* element capable of producing transposase. Equipped with sequence and map information of an *Ac* element in maize, geneticists have been able to clone important maize genes using the transposon-tagging approach. When a gene is inactivated by an *Ac* element, proof of which can be demonstrated by a low frequency of trait reversion, the *Ac* element can be used as a probe for 'fishing-out' the tagged gene.

A tagged gene can be cloned by performing Southern blot analysis with the *Ac* element used as probe, and any polymorphism represents an excision/insertion event. The technique is described more fully in the review by Walbot (1992). The *Ac* system has proven extremely useful for cloning maize genes for traits such as plant pigment genes (*bronze*; the anthocyanin pathway), starch biosynthesis (*opaque2*), and abscisic acid (AbA) insensitivity (McCarty *et al.*, 1989). Other maize transposable elements such as *Mutator* (*Mu*) and the *En/Spm* system have been used to clone maize genes, as has the *Tam* system in snapdragon. As a result, the search began for transposable element systems in other species so that a similar approach could be taken.

However, it was soon demonstrated that the well-characterized *Ac/Ds* system could be genetically engineered into other plant species including tobacco, *Arabidopsis*, tomato and petunia. It was subsequently shown that the *Ac* element acted in a similar manner in these heterologous species, undergoing germinal and somatic excision and insertion, and maintaining the propensity to reinsert at a linked site (usually within 4 cM). As a result, the frequency of mutation in the surrounding area approaches 1024. It was not until 1993 that a plant gene was cloned using the heterologous system, that of the *Ph6* gene in petunia, which is involved in floral pigmentation.

Using the transposon-tagging approach, a number of plant disease resistance genes (R genes) have now been cloned. The first of these was the tomato *Cf-9* gene, which confers resistance to race 9 of the fungal pathogen *Cladosporium fulvum*, causal agent of leaf mould. The gene was tagged using a modified *Ac/Ds* system. The system relied on the use of a stabilized *Ac* element (s*Ac*), which was effectively a disabled *Ac* element that nevertheless produced *Ac* transposase and hence enables *Ds* transposition. The two elements are maintained in separate lines, and hence are both stable until controlled crossing is performed. This gives a significant advantage in that *Ds* insertion mutants can be maintained indefinitely in the homozygous or hemizygous form, and revertants can be isolated by crossing lines with s*Ac* tomatoes.

The transposon-tagging approach used to isolate the *Cf-9* gene, which was known by phenotype only, did have considerable advantages in that much was known of the pathogen. The hypersensitive response elicited by the virulent strain fitted into the gene-for-gene hypersensitive response proposed by Flor (1971), with a dominant avirulence gene (*Avr9*) interacting with the *Cf-9* gene to form an incompatible or hypersensitive response. *Avr*9 had previously been cloned and fully characterized, and was known to specify a 28 amino acid peptide which alone could elicit a necrotic response in resistant tomatoes. Crosses were made to develop a transgenic tomato line homozygous for *Cf-9*, but heterozygous for *Ds*. Crossing was performed as outlined in Fig.. As can be seen, most seedlings died from systemic hypersensitive response, with *Avr9* and *Cf-9* present in all cells. The only survivors would be individuals where *Ds* had inserted into the *Cf-9* gene. Where this happened in plants carrying s*Ac*, the seedlings were variegated for the hypersensitive response (somatic excision), and in the absence of s*Ac*, the survivors were stable (germinal excision).

This work was a major undertaking, involving the germination of approximately 160 000 progeny, yielding 118 survivors, representing 63 independent mutations. After considerable effort to characterize these, the *Cf-9* gene was cloned and characterized, with an open reading frame encoding an 863 amino acid protein.

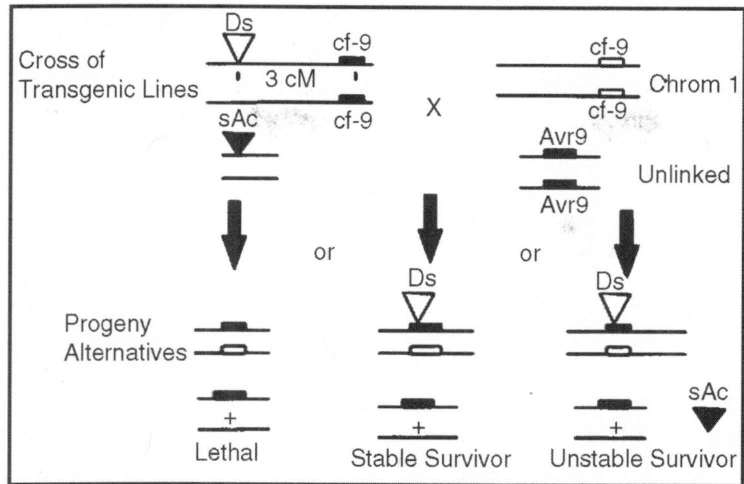

Fig. Graphical Representation of the Crosing Strategy of Transgenic Ac/Ds Tomato Lines.

Map-based Cloning

The previously described transposon-tagging scheme utilized to clone *Cf-9* had an element of the map-based approach by utilizing the phenomenon of *Ac/Ds* transposition to linked sites, with a *Ds* element within 3 cM of the locus of interest, which greatly enhanced the likelihood of insertional mutagenesis. However, map-based cloning requires a much tighter genetic linkage (in centimorgans or recombination fraction), which in turn needs to be converted into a physical distance (kbp of DNA sequence). The map-based, or positional cloning concept relies on building a well-saturated genetic map of DNA sequences of the organism in question. The genetic map is built up in a polymorphic mapping population, commonly an F2, backcross (BC) or recombinant inbred, with markers such as RFLPs (restriction fragment length polymorphisms). The first RFLP map to be made was of human. Based on such maps, important human genes responsible for disorders such as cystic fibrosis and muscular dystrophy have been cloned in this manner.

Genetic maps of plant species have been made with DNA markers in virtually all important plant species, particularly the major cereals, grain legumes, oilseeds and important vegetable crops including tomato and potato. The RFLP marker system is based on Southern blot analysis, while many of the newer technologies use the polymerase chain reaction (PCR). These include random amplified polymorphic DNA (RAPDs), amplified fragment length polymorphisms (AFLPs), and a number of different marker systems based on simple sequence repeats (SSRs), which may use PCR, such as inter simple sequence repeat (ISSR)-PCR, or Southern analysis. The applications of DNA markers to plant genetic analysis and breeding are many, and the subject has been reviewed by various authors.

The population for saturation mapping should be segregating in a Mendelian manner for the trait of interest. This requires a lot of tedious but demanding work, and it may be that particular populations such as near isogenic lines, or DNA pooling strategies, such as bulked segregant analysis, will allow the mapping effort to be more targeted to the region of interest, hence improving the rate of map saturation. It is desirable to obtain extremely close genetic linkage to the locus of interest (<1 cM), with flanking markers if possible. It can be of considerable advantage to identify a marker which co-segregates (say within 0.1 cM) with the trait. The next step is to convert the genetic map into a physical map of the chromosome region, translating distances from centimorgans to nucleotides.

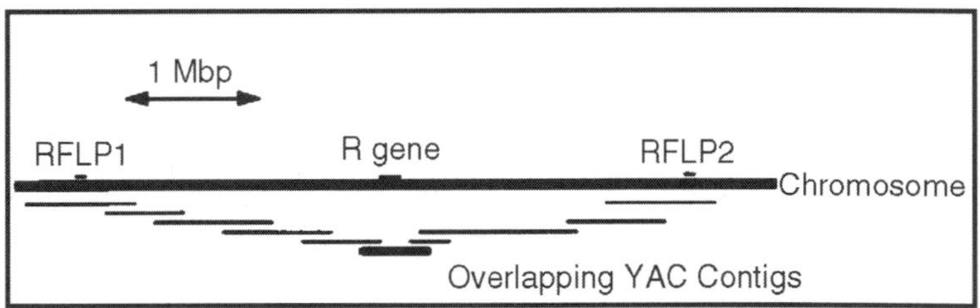

Fig. Generalized Strategy of Chromosome Walking Used in Map-based gene Cloning.

The Gene of Interest (in this illustration,R) is Flanked by Markers such as RFLPs or SSRs.A Series of Overlapping YAC Contigs is used to cover th Chromosomal Region Containing the Gene.The YAC Which Contains the Complete Gene is Marked in Bold.This is then Subcloned or Matched to a cDNA Library to Idenfify the Gene(s) Contained the Region, and Ultimately Identify the R Gene. Physical mapping then involves the use of YAC or BAC clones which span the region of interest. These YAC/BAC contigs will span the region between two flanking DNA marker sequences as shown in Fig., a process known as chromosome walking. In this example, two YAC/BAC clones contain the gene of interest. Methodology is being developed to enable BAC clones to be directly transformed into plants, to find which one contains the gene of interest (in the example, the *R* gene). Alternatively, the YAC/BAC must be subcloned (for example as a cDNA library), and these subclones can be put into the susceptible plant and tested for the R phenotype.

The first R gene to be cloned in plants via map-based cloning was the *Pto* gene, which confers resistance to *Pseudomonas syringae* pv. *lycopersici*, the causal agent of bacterial speck in tomato. The starting points for the work were:

- A high-density RFLP map of tomato.
- A tomato YAC library.

The *Pto* locus had been previously mapped to tomato chromosome 5 and one RFLP marker co-segregated with *Pto*. Hence this was a useful starting point to screen the YAC library. A 400 kb YAC was identified, and via inverse PCR, some end-specific probes were generated. The left arm of the YAC was 1.8 cM from *Pto*, with the right arm perfectly co-segregating. This YAC was then used to screen a 920 000 plaque leaf cDNA library, from which 30 were selected. One of these was shown to co-segregate with *Pto*. This cDNA was then transformed into a susceptible tomato cultivar under the control of the CaMV 35S promoter, where it conferred resistance. Hence the gene was identified, and could then be sequenced and characterized.

The success of this approach was partly due to the small genome size of tomato (950 Mbp), and a relatively low level of repeat sequence. Such an approach is much more difficult in species with larger genomes such as maize (2300–2700 Mbp) and wheat (16 000 Mbp). One such example of this problem is the 79 000 clone maize YAC library produced by Zeneca Seeds. As reported by K. Edwards, only 15% of the YACs have unique ends, with the remainder terminating in repeat sequences, which makes chromosome walking very difficult. The synteny of genomes is a major advantage in this instance. It is now well established that there is a high degree of synteny of genome organization, as seen in the co-linearity of maps of the most important grass genomes. This has resulted in the great interest in using this co-linearity to facilitate cloning of genes in large genomes such as maize and wheat, by starting with the smaller genomes, particularly rice, and to a lesser extent, sorghum. One particular study of maize YACs demonstrated that a single YAC with the *Adh1* sequence contained 36 different repetitive sequences, whereas these repetitive sequences were largely absent in corresponding areas of the rice and sorghum genomes, a pattern seen also for the *a1–sh2* region in all three genomes. In a similar manner, many genes of interest for *Brassica* improvement are being identified in the smaller *Arabidopsis thaliana* genome.

Paterson *et al.* (1995) have recently demonstrated the potential of this approach, by identifying quantitative trait loci (QTLs) for seed mass and phenology which were co-linear on molecular marker maps of maize, sorghum and rice. They proposed that, as the maize genome is four and six times larger than the sorghum and rice genomes, respectively, the genes of interest in maize may be cloned in sorghum/rice more expediently. It must be remembered that QTLs are subject to genotype 3 environment interaction, which will be a major complicating factor in attempts to clone and characterize a QTL.

T-DNA Tagging

When T-DNA is inserted into the plant genome, it does so via illegitimate recombination. In the same way as a transposon, the insertion of T-DNA into a gene will inactivate it, which phenotypically will appear as a mutation. The comparative advantage of such a method is that stable mutations will be conferred. This can be a disadvantage, however, as reversions, such as those made possible by the *Ac/Ds* system are not possible. It is also the case that each independent transformation event will result in just one mutation event (or more if multiple independent insertions are made), whereas the heterologous transposon approach allows multiple mutation events to take place from a single initial transformant. Hence, mutated genes with T-DNA insertions can be cloned, in much the same manner as those with transposon insertions, by using part of the T-DNA as a probe in Southern blots.

The procedure for selecting out the tagged gene is quite straightforward, particularly when the plant contains a single T-DNA copy. However, this means that a large number of independent transformants must be generated. This can be achieved with species such as *Arabidopsis*, where Feldmann (1991) demonstrated that a seed-based transformation system using *Agrobacterium* can be quite efficient. Feldmann (1991) demonstrated that by inoculating approximately 1000 seeds, approximately 300 000 selfed T2 seed can be collected. They estimated that approximately 8000 transformants could be recovered from an experiment of this scale and, using this approach, cloned a gene involved in trichome development and a dwarf gene from *Arabidopsis*.

A regulatory gene in the ethylene biosynthesis pathway of *Arabidopsis* has also been cloned using the T-DNA tagging strategy. With modification, T-DNA tagging methodology can be used to clone plant promoter sequences. Koncz *et al*. (1989) designed a vector which contained a promoterless antibiotic selection gene (aminoglycoside phosphotransferase II) and transformed this into *Nicotiana* and *Arabidopsis*. In this way, promoters could be 'trapped' as plants that were regenerated on selective antibiotic were antibiotic resistant because the T-DNA was inserted downstream of a promoter such that the gene was expressed.

The most serious limitation to the T-DNA tagging approach is that large numbers of independent transformation events must be generated, which is not currently possible with most species. This is really only feasible with a system such as the seed transformation technique developed for *Arabidopsis*, or possibly the 'whole-leaf ' transformation system developed for tobacco.

4

Forest Trees

BENEFITS OF URBAN FORESTS AND TREES

Trees in the urban environment contribute significantly to the aesthetic appeal of cities, helping to maintain the psychological health of the inhabitants. Besides the aesthetic and environmental aspects, urban forestry is also of importance in helping populations poor in resources to meet basic needs. Urban plantings of fruit trees to provide firewood is one way of utilizing obvious benefits from the available natural resources, as done in Kampala, the capital of Uganda.

Research has shown that urban trees benefit communities economically, socially and environmentally. The size, structure and condition of the urban forest directly affects the amount of benefits provided by the urban vegetation. Nevertheless, the social and environmental benefits provided by the urban forest are often largely ignored in land-use planning.

ECONOMIC IMPACT

Though trees are not usually thought of as economic resources, the presence of trees has an impact on different economic factors, including the level of real-estate prices, economic investments and employment. Results of recent Finnish studies show that the benefits provided by urban forests are reflected in property prices, and that environmental variables, such as proximity to wooded recreation areas and water courses as well as an increasing proportion of total forested area in the housing district, had a positive influence on apartment prices.

The presence of trees may influence property values positively by as much as 20%, with an average increase of 5-10%. Trees also provide different external environmental benefits. Costs for heating and cooling can be reduced by appropriate use of vegetation. Energy reductions for individual buildings have been shown to range from 5 to 15% for heating and 10-50% for cooling. Especially in areas with high summer temperatures, trees are important providers of shade. In areas with cool winters, it is of importance to use deciduous trees, as the loss of leaves in winter allows the sun to heat the house. Proper arrangements of vegetation around buildings can also reduce wind velocity and thereby reduce the heat loss from buildings.

DEFINITION OF TREES OUTSIDE THE FOREST

Trees outside the forest are defined by default, as all trees excluded from the definition of forest and other wooded lands. Trees outside the forest are located on "other lands", mostly on farmlands and built-up areas, both in rural and urban areas. A large number of TOF consist of planted or domesticated trees. TOF include trees in agroforestry systems, orchards and small woodlots. They may grow in meadows, pastoral areas and on farms, or along rivers, canals and roadsides, or in towns, gardens and parks. Some of the land use systems include alley cropping and shifting cultivation, permanent tree cover crops (*e.g.* coffee, cocoa), windbreaks, hedgerows, home gardens and fruit-tree plantations.

Classification of trees outside the forest presents certain difficulties. There are existing classifications for agroforestry, but none applicable to all trees outside the forest. For practical reasons, the FRA 2000 definition of "forest" combines aspects of both land cover and land use. This approach creates difficulties not only for classification of forest, but also for classification of TOF. In a study on data-gathering on TOF in Latin America, where classification was primarily based on land use criteria, separating land use and land cover aspects was found to be a main source of misinterpretation. There was a possibility of confounding coffee plantations and trees in pasture with forest, given their high density.

This clearly shows some of the problems involved in establishing a simple and reliable *a posteriori* classification.In France, the National Forest Inventory (IFN) and Teruti Land Use Study-have begun to attempt coordinating classifications of trees outside the forest. The objective is eventually to use the annual Teruti data to update the IFN ten-year data, with a single national nomenclature as a possible end result.

FUNCTIONS AND CHALLENGES

In industrialized countries, farmers list shade and shelter, soil protection and improvement of the landscape and rural environment as their main reasons for growing trees. In the tropics, farmers grow woody species for food security and subsistence. Trees outside the forest are a major source of food. Livestock fodder produced by TOF can be a matter of life and death in semi-arid or mountainous areas. Fuelwood remains the prime source of energy in developing countries, representing up to 81 per cent of the wood harvest (FAO 1999). In contrast, in the industrialized countries, fuelwood accounts for less than 10 per cent of total fuel consumption. Very few studies have reported on overall fuelwood output from stands and single trees outside the forest, but agroforestry systems and orchards are known to provide a large part of the resource. Trees outside the forest have an important ecological role. Planted trees and shrubs in fields help to check runoff and erosion and control flooding, as well as helping to purify water and protect against wind. Trees lining rivers and streams help to maintain biodiversity, providing spawning beds for fish and shellfish and shade which reduces eutrophication.

The unique role of trees in soil protection and conservation, checking wind and water erosion and maintaining soil fertility is universally acknowledged. Also important are the cumulative benefits of trees on smallholdings to soil and water conservation, in particular in the larger context of mountain watershed management; their positive impact on climate; and their role in buffering the effects of desertification and drought.

RESULTS OF SELECTED STUDIES

In spite of the limits of data at the regional or global level, a number of local initiatives have been carried out. The approaches of the various studies differ in accordance with the purpose and scale of the analysis. Few studies use methods resembling the conventional forest inventory. Many studies rely on existing literature or estimates drawn from surveys and interviews. The quantification of products is often based on different parameters, such as estimates of global output, marketed output, observed or potential productivity or economic value. Thus the reliability of the results is uncertain. The following are the results of some national initiatives that have assessed trees outside the forest.

In Kerala, the most densely inhabited state of India, a study estimated that of the total annual production of 14.6 million cubic metres of wood in the state, about 83 per cent was from homesteads (house compounds and farmlands), 10 per cent from estates (plantations of rubber, cardamom, coffee and tea) and only about 7 per cent from forest areas 26.6 per cent of the state area is under forest cover. Trees outside the forest met about 90 per cent of the fuelwood requirements of the state. Fuel from coconut trees alone, including both wood and non-wood materials (pruned and fallen), constituted about 70 per cent of the total fuelwood supply.

A study in Haryana State in India, an intensively cultivated state with about 3.8 per cent of its area classified as forest land but only about 2 per cent under actual forest cover showed that farm forestry (trees along farm bonds and in small patches up to 0.1 ha) accounted for 41.2 per cent of the total growing stock of wood. Multiple tree rows along roads and canals accounted for 13 per cent and 9.6 per cent, respectively; village woodlots for 24 per cent; and block plantations of less than 0.1 ha for 10.6 per cent. In Morocco, where forest cover is less than 5 per cent of the land cover and other wooded lands only 7 per cent, nearly 20 per cent of the land may be occupied by trees outside the forest, namely as wooded pasture (84 per cent) and fruit-tree plantations (12 per cent) (Rosaceae, citrus, olives trees, palm trees, walnut trees, fig trees, almond trees).

Fruit production has an important place in the national economy. It is noteworthy that even when a forest is largely destroyed, the carob is one of the few species traditionally conserved, as it is highly appreciated by farmers for multiple purposes, providing both fodder and income from the sale of its fruit for export. However, there are no reliable data on the distribution and potential of this "forest" resource, which is of interest for farmers, herders, concessionaires and the government and distributed on agricultural and forest lands.

In the Sudan, the National Forest Inventory has undertaken a national land use inventory to provide area and volume statistics for planning at the subnational and national levels. The inventory was designed to provide preliminary estimates regarding products other than the traditional fuelwood and timber, such as the amount of gum, fruit or nuts that can be collected and the distribution of non-wood species of interest.

In Costa Rica, the Tropical Agricultural Research and Higher Education Centre (CATIE), in collaboration with Freiburg University, Germany, is developing a regional methodology for Central America to assess tree resources outside the forest. A mix of satellite remote sensing, aerial photos and ground sampling is used to address the complexity of the resource and to allow dynamic monitoring of resources at the national and regional levels data-gathering on TOF in eight Latin American countries. None of these countries had established a database and the search for information was multisectoral. The statistics on land cover and land use gave some idea of the relative importance of trees outside the forest.

In Kenya, extensive tree planting on farmlands was promoted in the 1970s and 1980s, with land tenure security as a major incentive. There is an increasing trend of tree cover and species diversification on privately owned farms. Assuming that the present rate of increase in tree planting will continue, it was estimated that farms produced about 9.4 million cubic metres of wood in 2000 and will produce about 17.8 million cubic metres in 2020. Their share of the total wood produced in the medium- and high-potential districts was projected to increase to 80 per cent in 2020. Indeed, while natural stands of trees have declined there has been a corresponding increase in tree planting in much of the densely populated plateaus of Kenya. As natural forests are reduced or become inaccessible, agroforestry systems help people to diversify production and income and to protect themselves from shortages of fuel and wood.

In Bangladesh, natural forest formations cover less than 6 per cent of the country and the population growth rate is extremely high. An inventory of homestead/village forests in the country indicated that trees outside the forest constitute a vital resource for local populations, providing food, fodder and fuelwood. The sampling method was based on dual village/household sampling with an agro-ecological and administrative sampling base. Rural Bangladesh was divided into six major regions considered as agro-ecological strata, each subdivided into *thanas* (administrative entities, subdistricts). The households making up the sampling units were chosen at random from a number of villages.

The inventory sampled data on palm trees and cane as well as trees, bamboo and thickets. The results, expressed per stratum and per inhabitant, provide volumetric data for fuelwood and sawnwood and species data for total amounts under and over 20 cm. This inventory was apparently the first to nationwide assessment of trees outside the classified forests in Bangladesh.

RECREATIONAL USE OF GREEN AREAS

Several studies have shown the importance of urban forests for participation in outdoor recreation activities. In a Danish national survey it was found that around two-thirds of all forest visits take place in the forest situated nearest the housing area. Danish and Swedish studies have also showed that the use of green areas by people is associated with three highly esteemed values: (i) a high degree of natural elements, (ii) fresh air and (iii) the possibility of solitude.

In general, people use green areas less than they would like. According to Swedish studies, distance and the fear of assault are the most prevalent reasons for people not visiting green areas. When the distance to the park exceeds 300 m, one person in four postpones a daily visit. As many as 56% refrain from regular walks in the park when the distance increases to 500m. American studies indicate that the removal of vegetation at strategic points can reduce the risk of assault, although this should not be the reason for removing all the vegetation.

PSYCHOLOGICAL ASPECTS

City life is stressful but research shows that urban green areas have a beneficial influence on the health and well-being of the urban population. Studies indicate that visits to green areas can counteract stress, renew vital energy and speed healing processes. In Sweden, Grahn (1989) conducted extensive studies on the significance of parks to different groups of the population. For instance, he persuaded 40 schools, hospitals, sports associations, cultural associations and day-care centres to keep diaries of their outdoor activities for 1 year. For example, the diaries show that periods spent outside had an actual medicinal value for patients and residents of hospitals, old people's homes and homes for the sick. People became happier, slept better, needed less medication, were less restless and far more talkative.

Ulrich (1984) showed that hospitalized patients recovered faster when they had a view through a window, allowing them to see trees. Ulrich *et al.* (1991) showed a gory film on industrial accidents to 120 people. Half the people were then shown a nature film, whereas the other half were shown a film on the city, with sequences of buildings and traffic. The subjects' heart beat, muscular tension and blood pressure were monitored throughout. All subjects exhibited strong signs of stress during the first film on industrial accidents. The stress levels of the subjects that then watched the nature film returned to a normal level after 4-6 min, whereas the half that watched the film on buildings and traffic continued to exhibit high stress levels.

Kaplan and Kaplan (1989) have formulated a theory on the interaction between human attention and the surroundings. This theory distinguishes between spontaneous attention and conscious attention. Spontaneous attention demands no effort and occurs without premeditation. Conscious attention demands energy and leads to psychological exhaustion in the long term. Rapid movements, strong colours, sudden noises and strong odours are typical stimuli that demand conscious attention. These signals have a powerful effect on our attention, consciously and unconsciously, because they are perceived as potential dangers to which we should react. This means that urban living, with fast vehicles, flashing neon signs and strong colours, causes constant stress.

Kaplan and Kaplan's research indicates that vegetation and nature reinforce our spontaneous attention, allow our sensory apparatus to relax and infuse us with fresh energy. Visits to green areas bring relaxation and sharpen our concentration, since we only need to use our spontaneous attention. At the same time, we get fresh air and sunlight, which have significance for our diurnal and annual rhythms.

Environmental Education

The change and the continuity in nature provides not only a sense of time but also a sense of security and confidence, through the predictable change of seasons and the repetition of natural processes. The urban trees and forests play a role in educating young people and children in understanding the basic processes in nature and the complexities of the environment, either informally during play or as part of the school curriculum. Playing in nature or natural settings also helps children to develop their motoric senses and to regain and experience the human connection to nature through their imagination. Experiences with environmental education in the USA show that access to even a small area of the urban forest is of importance when learning about natural processes.

In England in 1985 a research project supported by the Department of Education and Science, the Countryside Commission and a consortium of local authorities resulted in the organization Learning through Landscapes (LTL). LTL was founded on the recognition of the impact that the physical surroundings has on children. LTL helps schools to design schoolyards that improve the quality of the environment and helps to create additional resources for the formal curriculum, such as the promotion of more effective teaching and learning in the outdoor classroom. The English Community Forests are examples of integrated programmes in which environmental education is integrated in social programmes by well-coordinated planning. The forests are charged with running and coordinating social programmes, involving information, consultation, participation, art, education, sport, recreation and cultural activities.

Community Involvement

Other benefits of urban forestry are those provided by involving people in planning, planting and caring for the trees and urban forests in their own locality. By promoting social interaction and strengthening local pride and identity, local public engagement benefits management greatly, encouraging people to act protectively towards their local forests and providing a potentially huge volunteer workforce.

The possibilities for creating public awareness in any town are endless. In the USA there is a tradition of involving the public in all processes, from the policy-making to the care and maintenance of the urban forest. Through various types of tree programmes, public awareness has proved to be a valuable tool in the justification of tree activities because it develops a public interest group that helps to lobby for support. The projects that involve citizens have various aspects in common, such as planting, preserving, pruning, maintenance, public education, etc..

Examples of initiatives that enhance public awareness and engagement in a municipal tree programme are the Tree Boards, municipal non-profit groups with participants from a wide range of professions. The close cooperation between professionals and non-professionals provides direct communication of new ideas and research and helps to solve associated problems as experienced in a practical situation. Another way of involving the public is to encourage houseowners to adopt a tree, by taking part in the care and maintenance of a tree in their neighbourhood.

THE URBAN ENVIRONMENT AND FOREST TREES

Trees in the urban environment are often referred to as the urban forest, comprising trees in civic woodlands, parks and the street. Earlier, urban trees were mainly regarded as aesthetic elements, whereas today they are recognized as having a positive impact on the environment as well as providing economic and social benefits. Monetary evaluations reflect the various benefits arising from the urban forest, covering such aspects as reduction of pollution and energy use, environmental amelioration, savings in public health care and increase in economic investment. Hence, the value of the urban forest is being increasingly recognized as a vital component in the maintenance of a sustainable urban environment in cities around the world. Meanwhile, the population living in urban areas has increased rapidly since the 1950s and the lack of space makes it tempting to use green areas for infrastructure and buildings.

Collins (1997) emphasizes that urban forestry, the planning, management and maintenance of the urban forest, is more closely aligned to traditional forestry than might be immediately apparent. As implicit in the term, the principle of sustained yield has been adapted to the urban environment, applying rural land-use forestry to the management of the urban forest. Hence, the overall objective in urban forestry is not that of timber production but, through a balanced structure of age and species, a sustained production of environmental, social and economic benefits. These social benefits also accrue if local communities are encouraged to contribute to their own environment by promoting projects and activities involving local residents. These kinds of projects often prove effective in promoting social interaction and lead to increased local involvement in other aspects as well.

The planning and management of a healthy urban forest requires that coordinated strategies are agreed between the professions more or less directly involved, such as planners, landscape architects, arboriculturists, foresters, engineers, legislators, developers and utility managers. Management of the urban forest resource is moving towards the achievement of a healthy, well-distributed urban forest, which is increasingly becoming an essential and integral part of the urban infrastructure.

Trees in the urban environment have been defined in several ways, although 'urban forestry and arboriculture' is probably the most used term in relation to trees in or near the urban environment. Many different research disciplines are involved in the fields of arboriculture and urban forestry. Harris (1992) defines arboriculture as being 'primarily concerned with the planting and care of trees and more peripherally concerned with shrubs and woody vines and groundcover plants'. However, many people consider urban forestry to consist primarily of two types of planting: urban forests and urban trees. Also, definitions of urban green areas and urban forestry vary significantly throughout the world. There seem to be different opinions as to what urban forestry covers, depending on whether professionals have a background within or outside forestry or have experience of working in the USA or other parts of the world such as Europe.

In many cases foresters have argued that urban forestry concerns 'forestry in urban areas'. Volk (1986) stated that: Green areas such as tree lined streets, cemeteries, and parks, which are not predominated by trees and as such do not come under forestry influence should be excluded from consideration [of being considered urban forestry]. On the other hand, park forests which come under forestry supervision, should be included in our considerations. The fear of involvement of foresters in urban plantings has been described by Chambers (1987): 'Misguided resistance to the concept [of urban forestry] is still often mistakenly assumed to imply the take over of public open space and existing amenity trees for timber production.' Americans tends to look at urban forestry as the 'management of trees in urban areas on larger than an individual basis'. In Europe this broad concept now seems to be accepted on a wider basis. Urban forest stands are now often considered as resources where an economic yield is not required, and traditional forestry practices are combined with both aesthetic and recreational considerations.

Costello (1993) suggested that urban forestry be defined as 'the management of trees in urban areas', including single trees. In this definition, 'management' is described as the planning, planting and care of trees; 'trees' as individuals, small groups, larger stands (*e.g.* green belts) and remnant forests; and 'urban areas' as those areas where people live and work. The location of urban forests can be anywhere, from urban settings to the countryside, as long as there are human structures on the site; these structures can be related to ecological functions, protection, merchandise or tourism and recreation.

Using this concept, we categorize trees in the urban environment as both urban forests and urban trees. Urban forests can be defined by their placement in or near urban areas and by their multifunctional aspects, giving shade, amenity values, etc. Therefore, we define urban forestry as the establishment, management, planning and design of trees and forest stands with amenity values, situated in or near urban areas. Trees in the urban environment may be divided into three different types: trees in urban woodlands, park trees and street trees. These types of plantings differ distinctively in several ways. Most importantly, they have significantly different growing conditions, which means that they have different needs with regard to planning, management and maintenance. The methods used and the research problems related to urban green areas are common throughout the world due to the fact that urban 'greening' concerns not only natural spheres (such as trees, growing conditions, etc.) but also the environment in close proximity to human populations.

URBAN GROWING CONDITIONS

Urban growing conditions differ significantly from those in the rural landscape, and produce difficulties as a result of both above- and below-ground influences.

Stress Factors

The harsh soil and air conditions that exist in urban planting are problems that do not play the same role in landscape planting. Growing conditions may also be difficult due to shading effects, recreational users, etc.. The modified urban mesoclimate affects the quantity of contaminants in urban areas, which is raised by a factor of around 25. In general, the average lifespan of a newly planted street tree may be as low as 10-15 years.

During the last 30-40 years, the vitality of street trees has fallen drastically. Heavier traffic patterns have increased demands for road construction, which consequently has changed the growing conditions of many roadside trees. Also, pollution from traffic has a highly detrimental impact on street trees. The fact that 50% of the trees planted in an urban environment die within the first year emphasizes this point. Nowak *et al.* (1990) found that 34% of 480 trees died within 2 years of planting, while Miller and Miller (1991) found that the mortality rate was 25-50% for a number of species planted in Wisconsin, USA.

Temperature extremes can occur, especially where trees are widely spaced and where heat is reflected from hard surfaces. Harris (1992) described that, occasionally, tree limbs up to 0.6 m and trunks up to 1.2 m in diameter break and fall during hot calm summer and autumn afternoons and subsequent evenings. Roots are more sensitive to temperature extremes than the tops of plants.

Wind speed will vary according to the shape and height of buildings. Areas with tall buildings will usually be relatively cool in summer due to shading effects, and warmer in winter due to wind-protection effects. On the other hand, winds are more variable and more extreme at exposed corners of tall isolated buildings. Buildings deflect strong winds downwards and concentrate their force at the base and corners of buildings, forming 'wind tunnels'. Trees planted in these exposed gaps may suffer scorched leaves and shoots, which lead to a stunted canopy, especially on the windward side. Newly planted trees will transpire more rapidly in windy situations, which can lead to the death of a tree already severely stressed by drought. The wind stability of trees is determined by tree species, stand structure, spacing, thinning regimes, soil classes, breeding and tree age at the time of anchorage.

The presence of airborne pollutants in the atmosphere has been a characteristic feature of the urban environment since the beginning of the Industrial Revolution. Air pollution can occur in a variety of forms but the principal ones are dust, SO_2 and NO_x. Leaves are the plant parts most likely to show symptoms of air pollution injury. On broadleaved plants, leaves may develop interveinal necrotic areas, marginal or tip necrosis, stippling of the upper surface, or silvering of the lower surface.

However, trees in the urban environment also play a role in the quest for cleaner air in the cities. Scott *et al.* (1998) showed that daily uptake of NO_2 and particulate matter represented 1-2% of anthropogenic emissions for the county of Sacramento, California. In areas with winter temperatures below 0°C, the use of de-icing salt is a well-known problem. De-icing salt is applied to the surrounding environment by surface run-off, wet spraying and airborne drifting.

The initial and most common symptom of de-icing salt damage on trees and shrubs is reduced growth. This is often difficult to recognize or may be confused with other stress factors. Reduced growth is usually followed by early autumn colours and premature leaf fall. De-icing salt is usually accumulated on the windward side of trees. The damage is easily recognized because it faces the road and is normally regarded as the best indication of de-icing spray damage. The majority of trees and shrubs subjected to either soil salt or salt spray typically show necroses at the edges of the leaves or needles. Wounds, often related to pruning, are a common place for spray salt to infect the plants. The damage may cause lack of sprouting and eventually dieback. Conifers are very susceptible to de-icing salt spray damage because they are green all through the winter maintenance season. Trees and shrubs damaged by de-icing salt and showing dieback are difficult to cure.

Characteristics and Restriction of Rooting in the Built Environment

Urban soils as a growing medium are poorly understood and often misunderstood. Therefore plantings are carried out with little appreciation or attention to the character and quality of the material that lies beneath the surface. One major problem in relation to planting in the urban situation is soil compaction, which may occur in small as well as large urban sites. Soil compaction can be divided into two types: (i) intentional soil compaction, which occurs when soil is deliberately compacted for site stabilization under roads, houses, etc. and (ii) unintentional soil compaction, which occurs when traffic uses areas intended for planting (Randrup 1997). In the urban situation, unintentional soil compaction is primarily found along roadsides and on construction sites.

When soil is compacted, its bulk density increases and its porosity decreases. These effects inhibit plant growth because the soil becomes impenetrable to root growth and, furthermore, restricts the water and oxygen available to the roots. For example, root growth of most plants is impeded once soil bulk density rises above 1.6. One consequence of compacted soil is waterlogging, which can kill roots around existing trees. Soil loosening has proved to be effective in alleviating compacted soil. However, there is no doubt that the best treatment for compacted soil is to protect the soil from being compacted in the first place. Florgård (1987) suggested protecting trees growing on construction sites by dividing the site into zones in which different types of construction traffic are permitted. The principle of construction site zoning was adapted by Randrup and Dralle (1997) to protect the soil from being compacted. They suggested that the entire construction site be divided into a building zone, a working zone and a protection zone.

URBAN CHARACTERISTICS/ENVIRONMENT ANALYSED

The city is characterized by paving and buildings, which variously results in decreased wind speed, but an increase in windthrow, raised temperatures and precipitation, lowered humidity and shading in street canyons. The extent of the influences of these factors depends on the structure and amount of vegetation, as well as on the size of the city. Green areas in the city accessible to the public include areas at schools, public libraries and social institutions, as well as parks and churchyards. These are areas that are typically relatively small and geographically widespread, which is why they often suffer great recreational pressure, most often due to the overall proportion of green areas in the city.

Throughout Europe the proportion of urban green areas varies greatly, from over 60% of the area of Bratislava, the capital of Slovakia, to about 5% in Madrid, the capital of Spain. In comparison, the figure for Mexico City is only 2.2%; in relation to the number of inhabitants, this only provides 1.94 m^2 per inhabitant, which is far below the 9 m^2 recommended by the World Health Organization. A suggested measure of urban environmental quality is the location of green areas within a walking distance of 15 min or less from all housing areas. This criterion is met for all citizens in Brussels, Copenhagen, Glasgow, Gothenburg, Madrid, Milan and Paris and, in general, for more than 50% of the population in most European cities.

The quality of urban green areas are increasingly recognized as being important to the overall quality of human life in the cities. Beyond this, the urban forests and trees are important as ecosystems in relation to the conservation of biological diversity. Though the urban population benefits from the urban green areas, the increase in population places great pressure on the existing green areas as a result of urban and infrastructural development. The importance of the urban green space, of which urban forestry is an integral part, increases as population increases.

The growing urban population needs the environmental and social benefits associated with urban forests, as shown by Ulrich (1984) and Grahn (1989). In developing countries, urbanization has had a dramatic influence on creating environments practically without any amenities, as is the case in Mexico City, where the growth in population has not been matched by an increase in green space. It is an accepted reality that the growth of the cities cannot be stopped. Instead, the challenge is to control urban growth so that it results in economic growth and a satisfactory environment.

As more than two-thirds of the population of Europe live in urban areas, the quality of the urban environment, including green areas, is becoming increasingly recognized as one key to the economic reconstruction of European cities (Commission of the European Community 1990). Urban areas constitute the everyday environment of the greater part of the population and, in recent years, this topic has received considerable attention in the European Union, United Nations and the Organization for Economic Cooperation and Development, which will hopefully lead to an improvement in our knowledge of urban environments.

FRUIT TREE PRUNING

Pruning fruit trees is a technique that is employed by gardeners to control growth, remove dead or diseased wood or stimulate the formation of flowers and fruit buds. The most economical pruning is done early in the season, when buds begin to break, and one can pinch off the soft tissue with one's fingers (hence the expression "nipped in the bud"). Many home fruit growers make the mistake of planting a tree, then neglecting it until it begins to bear. But careful attention to pruning and training young trees will ultimately determine their productivity and longevity. Good pruning and training will also prevent later injury from weak crotches that break under snow or fruit load.

OVERVIEW

To obtain a better understanding of how to prune plants properly, it is useful to have some underlying knowledge of how pruning works, and how it affects the way in which plants grow. Plants form new tissue in an area called the meristem, located near the tips of roots and shoots, where active cell division takes place. Meristem growth is aimed at ensuring that leaves are quickly elevated into sunlight, and that roots are able to penetrate deeply into the soil. Once adequate height and length is achieved by the stems and roots, they will begin to thicken to give support to the plant. On the shoots, these growing tips of the plant are known as apical buds.

The apical meristem (or tip) produces the growth hormone auxin, which not only promotes cell division, but also diffuses downwards and inhibits the development of lateral bud growth which would otherwise compete with the apical tip for light and nutrients. Removing the apical tip and its suppressive hormone allows the lower dormant lateral buds to develop, and the buds between the leaf stalk and stem produce new shoots which compete to become the lead growth.

Manipulating this natural response to damage (known as the principle of apical dominance) by processes such as pruning (as well as coppicing and pollarding) allows the horticulturist to determine the shape, size and productivity of many fruiting trees and bushes. The main aim when pruning fruit trees is usually to obtain a decent crop of fruit rather than a tree with an abundance of lush yet unproductive foliage. Unpruned trees tend to produce large crops of small, worthless fruit often damaged by pests and diseases, and much of the crop is out of reach at the top of the tree.

Branches can become broken by the weight of the crop, and the cropping may become biennial (that is, only bearing fruit every other year). Overpruned trees on the other hand tend to produce light crops of large, flavourless fruit that does not store well. Pruning is therefore carried out to achieve a balance between shoot growth and fruit production.

Formative Pruning of Bush Trees

Formative pruning of apple and pear trees (the pome fruits; the stone fruits such as cherries, plums, gages, etc., have different requirements and should not be pruned during the dormant months) should be carried out during the dormant winter months between November and March (or June and September in the southern hemisphere) and during the early years of the tree's life to develop a strong framework capable of bearing the weight of the crops that will be borne in later years. This involves hard pruning, although in later years pruning will be lighter and carried out to encourage fruiting.

One year old tree

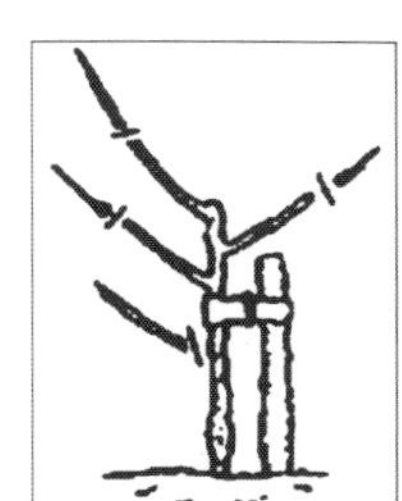

Two year old tree

Three year old tree

Four year old tree

Maiden Tree

A maiden whip (that is, a one year old tree with no side shoots) should be pruned to a bud with two buds below it at about 80 cm from the ground immediately after planting to produce primary branches during the first growing season.

A feathered maiden (that is, a one year old tree with several side branches) should have its main stem pruned back to three or four strong shoots at 80 cm from the ground. Side shoots should be shortened by two thirds of their length to an upward or outward facing bud. Lower shoots should be removed flush with the stem.

Two Year

Remove any lower shoots and prune between three and five of the best placed shoots by half to an upwards or outwards facing bud to form what will become the tree's main structural branches. Remove any inwards facing shoots.

Three Year

Prune the leading shoots of branches selected to extend the framework by half to a bud facing in the desired direction. Select four good laterals to fill the framework and shorten these by a half. Prune any remaining laterals to four buds to form fruiting spurs.

Four Year

The tree will have begun to fruit and only limited formative pruning is now required. Shorten leaders by one third and prune laterals not required to extend the framework to four buds.

Five year and Onwards

The tree is considered to be established and should be annually pruned as described in

Pruning the Cropping Tree

Before pruning it is important to distinguish between spur bearing and tip bearing varieties. The former, which is the most common type, bear most of their fruit on older wood, and include apples such as *Cox's Orange Pippin*, *James Grieve* and *Sunset*, and pears such as *Conference*, *Doyenne du Commice* and *Williams Bon Chretien*. Tip bearers on the other hand produce most of their fruit buds at the tips of slender shoots grown the previous summer, and include the apples *Worcester Pearmain* and *Irish Peach*, and the pears such as *Jargonelle* and *Josephine de Malines*. There are basically three types of pruning that are applied once the main shape of the tree has been established.

Renewal Pruning

- Spur pruning:Spur bearing varieties form spurs naturally, but spur growth can also be induced.
- Renewal pruning: This also depends on the tendency of many apple and pear trees to form flower buds on unpruned two year old laterals. It is a technique best utilised for the strong laterals on the outer part of the tree where there is room for such growth. Pruning long neglected fruit trees is a task that should be undertaken over a lengthy period, with not more than one third of the branches that require removal being taken each year.
- Regulatory pruning: This is carried out on the tree as a whole, and is aimed at keeping the tree and its environment healthy, eg, by keeping the centre open so that air can circulate, removing dead or diseased wood, preventing branches from becoming over crowded (branches should be roughly 50 cm apart and spurs not less than 25 cm apart along the branch framework), and preventing any branches from crossing.

Pruning of tip Bearers

Tip bearers should be pruned lightly in winter using the regulatory system. any maiden shoots less than 25 cm in length should be left untouched as they have fruit buds at their tips. Longer shoots are spur pruned to prevent over-crowding and to stimulate the production of more short tip bearing shoots the following year. Branch leaders are 'tipped', removing the top three or four buds to a bud facing in the desired direction to make them branch out and so produce more tip bearing shoots.

5

Applications in Forest Tree Improvement and Breeding

Three broad areas of application of cryopreservation and *in vitro* storage techniques to forest tree improvement warrant discussion:

GERMPLASM PRESERVATION

In vitro storage is now routinely used for germplasm storage of cassava, at CIAT (Escobar *et al.* 1992, Chavez *et al.* and for many *Solanum* genotypes at CIP, in Peru (Withers *et al.* 1990). ORSTOM, at Montpellier, is developing methods for the preservation of coffee by *in vitro* culture of immature zygotic embryos (Charrier *et al.* 1991). Work aimed at the storage of somatic embryos of mango is underway (IBPGR 1991). In the IBPGR *in vitro* program, high priority is being given to the development of good *in vitro* culture procedures for sweet potato, taro, cocoyam, banana and plantain, sugarcane, cocoa, citrus and forage grasses (Withers *et al.* 1990). With respect to cryopreservation, preservation of a substantial number of genotypes of oil palm by ORSTOM is the closest to routine application (Engelmann 1991), while the development of routine procedures for banana and plantain (at CATIE), cassava (at CIAT) and coconut is underway (IBPGR 1991).

A number of features characterize the above crops targeted for *in vitro* or cryopreservation approaches to gene conservation:

- They are limited in number, and are all of major, widespread importance.
- Most are characterized by a long history of domestication and use.
- Biological and agronomic features are relatively well known.
- Several are vegetatively propagated.
- Some have recalcitrant seeds.

Gene conservation in these crops is to some extent then directed towards:

- The collection of known genotypes from many different areas of use. These genotypes have been evaluated to some degree, and it is desirable gene combinations, rather than individual genes, which are the target of preservation.

- Preservation of threatened wild relatives, characteristics of which would also be partially known.

Under these circumstances, conservation priorities are reasonably easily set. For these crops, *in vitro* storage and cryopreservation offer alternatives to the traditional dependence on field genebanks. For agricultural crops where seed collection and storage are possible, however, seed storage is the recommended approach to gene conservation (Withers 1992). Undoubtedly, urgent gene conservation and preservation problems exist for forest trees species, mainly among the tropical hardwoods and non-industrial species. Compared to the crop species discussed above, the following features characterize these tree species:

- There is a large number of species of known or potential interest. Many may never be used commercially, and some will be important only locally.
- Biological features of many are poorly known.
- Species distributions, and those of relatives, are poorly known.
- Seeds of some species will be in the recalcitrant category, but many are orthodox.

Major impediments to the preservation of these gene pools are:

- Resources available are sufficient for only a very small fraction of the survey and collection work which would be required.
- Seed storage facilities in many tropical countries are unreliable, such that satisfactory storage of seed collected now cannot be guaranteed (Ng 1985, Reid *et al.* 1991).

Wang *et al.* (1993) correctly drew attention to the potential use of *in vitro* storage and cryopreservation as a complementary method for the *ex situ* conservation of long-lived perennial species. This potential is undoubtedly real. It is of value though to carry this a little further, to consider specifically to which groups of tree species the technology might be applicable in the near future. It can be observed that replacement of existing seed storage facilities with more sophisticated technology is not likely to make a positive impact on the gene conservation problem with tropical tree species. Even for the recalcitrant species, activities should favour the establishment of *ex situ* plantings which will enable evaluation of the material. Cryopreservation and *in vitro* storage techniques thus offer no solution to erosion of tropical forest gene pools. More concerted efforts at documentation and *in situ* conservation are required.

Of established industrial species, cryopreservation and *in vitro* storage of vegetative material are not advantageous approaches to gene conservation in most of the conifers and eucalypts, for which seed can be stored satisfactorily. Poplars display some of the features characteristic of the fruit tree species described above - problematic seed storage, and breeding and commercial use based substantially on known clones. *In vitro* storage or cryopreservation may have a role as a back-up measure in large poplar breeding programmes. Similarly also for other species with seed production and storage problems, e.g. *Juglans*. It should be noted though that, for most forest tree species, introduction of *in vitro* or cryopreserved material into a breeding program will involve a delay of several years to flowering, and generally also for evaluation. Flowering genotypes in field clone banks can be much more quickly incorporated. An exception to the flowering obstacle is stored pollen, which can be rapidly incorporated into breeding programmes, and pollen storage should be more prominent in gene conservation programmes with industrial species. Cryopreservation may facilitate the broader use of pollen storage in these programmes.

It should be noted also that, as pointed out by Jackson and Coleman (1991) and Wang *et al.* (1993), the use of cryopreservation and *in vitro* storage could involve selection for genotypes which tolerate the conditions imposed (e.g. freezing and thawing), and which are capable of regeneration following such treatment. The conserved material is thus not necessarily a random sample of the genetic material collected.

Maintenance of Juvenility

Cryopreservation is perhaps the most promising approach currently on offer for maintenance of the juvenile state and capture of genetic gains through clonal forestry with industrial species (Haines 1992). The technology is applicable mainly in cases where good breeding programmes are in place and clonal forestry is a realistic goal. In particular, this applies to some programmes with conifers. Although some good breeding programmes are in place for the eucalypts, the need is not so urgent for these because rejuvenation can be achieved by other means. The approach ultimately may have some application to other tropical hardwoods, as good breeding programmes are established for these. Maturation is a much less important issue with many of the non-industrials, in particular those which flower precociously, can be selected early, and which can still be readily propagated at selection age.

Transport of Germplasm

The use of plants, organs or cells cultured *in vitro* circumvents to some extent the limitations which quarantine regulations place on the import of vegetative material (in particular) into many countries. International germplasm exchanges are becoming very important in the breeding of many crops, in particular to and from international genebanks at the International Agricultural Research Centres (IARCs) and other centres. *In vitro* exchanges are becoming common for some of these, e.g. for potato, coconut, banana and oil palm (Charrier *et al.* 1991, Withers 1992). In addition, *in vitro* collection, overcoming problems of inadequate or immature seeds and deterioration of vegetative material in transit, is now being used. Simple field procedures have been devised for cocoa, avocado, coconut, citrus, cassava, cotton, and forage grasses. (Withers *et al.* 1990, Withers 1992).

With the increasing emphasis on cooperative breeding of both industrial and non-industrial forest tree species in tropical countries, international movement of material will become increasingly important. In the short term, such movement is likely to be mainly in the form of seed for species and provenance testing. Movement of vegetative material, however, may become more important as breeding programmes advance, and *in vitro* approaches could find application in the longer term.

IMPROVEMENT OF TREE BREEDING

Breeding involves crossing among selected trees to produce progenies with new combinations of genes for testing and selection. The resulting selections will be the clones of the next generation of improvement. Breeding requires flowering trees of the selected clones that will be used as parents. This is a major problem for exotics like P. deltoides, since almost no provisions have been made to establish and maintain breeding orchards of mature individuals of the introduced clones.

All controlled crosses involving P. deltoides before 1997 have been opportunistic, depending on what was flowering. Seven P. deltoides clones flowered during this time (G-48, D-121, and S7C8 were females, and G-3, S7C1, S7C15, and S7C20 were males). Except for D-121, all of these were from Brazos County, Texas. One male flowering clone of P. yunnanensis and one male of P. `robusta' (Euramerican hybrid) were also used. "Breeding" in U.P. has been by the U.P. Forestry Department's Lalkuan Research Centre, Silviculture Division, Sal Region, by nearby WIMCO Seedlings Ltd. in Rudrapur, and at FRI in the Genetics and Tree Propagation Division. Initially, most of the new clones came from open pollinations between G-48 and G-3 (the first two clones to flower).

These seedling progeny clones provided the "L-series" clones of the U.P. Forestry Department and most of the "WSL-series" clones of WIMCO. Subsequently, the U.P. Forestry Department nursery manager at Lalkuan made the following crosses (female x male): G-48 x G-3, D-121 x S7C1, G-48 x P. ciliata, D-121 x P. ciliata, G-48 x P. yunnanensis, and G-48 x P. `robusta'. WIMCO selected three seedling clones from their own controlled crosses of G-48 x G-3, and they made some backcrosses (such as G-48 x P. `robusta) to stabilize desired hybrid combinations of P. deltoides with P. nigra. Controlled crosses by the Noh, E.R., K.K.Sharma, and M.L.Kapoor. 1995. A report submitted to FAO on Improvement Programme of Indigenous Poplars with Particular Reference to India. FRI, Genetics & Tree Propagation Division, Dehradun.

Genetics and Tree Propagation Division at FRI were made in the spring of 1996 and required transport of flowering branches from Lalkuan. The crosses G-48 x G-3 and G-48 x P. ciliata were successfully accomplished. The P. ciliata male flowers came from the hills of nearby Mussoorie.

The greatest amount of controlled crosses among Populus species in India has taken place at the Y. S. Parmar University of Horticulture and Forestry in Solan, H.P. Flowers (male and female) of P. ciliata could only be successfully ripened on cuttings in the greenhouse if the cuttings were cleft-grafted to potted rootstock seedlings in January before bud break. Water cultures and 'twig in pot' methods did not work. Cuttings of P. gamblei bearing male flowers could not be forced to ripen and shed pollen at Solan under any method, so no crosses with this species were made. Dr. Khurana successfully produced the hybrids P. ciliata x P. maximowiczii (pollen obtained from Japan), P. ciliata x P. yunnanensis, and P. deltoides (female) x P. ciliata (male) (but not the reciprocal) (personal communication). The P. ciliata x P. maximowiczii hybrid was heterotic for growth, but had the branch knot problem of P. maximowiczii. Dr. Khurana selected 20 clones from field tests of this hybrid for minimum branch knots.The P. ciliata x P. yunnanensis and P. deltoides x P. ciliata hybrids did not show heterosis for growth in his tests.

Backcrosses of the Euramerican hybrid clone '1-455' (female), which exhibits fast growth but is highly susceptible to Melampsora spp. of leaf rust, with P. deltoides (male) by Dr. Khurana produced two selections that exhibited extremely fast growth (4 to 5 meters per year in height) on recently-exposed, well-drained forest sites (not agroforestry) in H.P. A total of 358 clones (primarily P. deltoides, including some intraspecific crosses of G-48 by G-3) have been preserved in a germplasm bank at FRI in Dehradun, U.P. There are also 277 clones of P. deltoides in a germplasm bank at the HFRI Shilly Research Nursery in Solan, H.P.

Most of these are duplications of the clones at FRI, so that protection against a catastrophic loss of clones (from fire, etc.) is provided. The U.P. Forest Department's Lalkuan Research Centre is also maintaining a collection of clones from previous introductions and from inter-and intraspecific crosses. Many of these are duplications of clones at FRI. There are 150 clones in the Department's nursery at Lalkuan (100 P. deltoides clones), another 60 clones in a replicated test planted in February 1985 at `Ganga Pur Patia East' near Lalkuan, and 233 clones in the Department's populetum planted in February 1989 at Tanda, U.P. (Plot #47). Forty-four of the clones in the Tanda planting came from Dr. Hansen's 1986 shipment of 200 clones to India and were selected based on nursery performance at the Lalkuan nursery. The Ganga Pur and Tanda tests were just beginning to flower in 1996 and will provide the only breeding orchard available in India (108 clones) for the initiation of an advanced-generation mating design with P. deltoides.

TREE IMPROVEMENT

The long history of selection and improvement of many agricultural crops contrasts with that for forest tree species, where the shift from exploitation to domestication is much more recent. It is generally recognised that sound management of genetic resources is critical to the optimisation of reforestation efforts - for both industrial and non-industrial applications.

For industrial forestry species, of which there are an estimated 100 million hectares of plantations worldwide, the most common approach to improvement is recurrent selection in genetically quite broad breeding populations. Four generations of breeding have been completed in *Eucalyptus grandis* and three in some *Pinus* species, but very few programmes are this advanced. The use of open-pollinated general combiner seed orchards is the most common approach to the capture of genetic gain, while a few advanced programmes are characterised by deployment of superior full-sib families or clones. Reproductive biology and patterns of variation are reasonably well documented for the established industrial species. Vigour, form and wood quality are the priority selection criteria in most programmes, with disease or insect resistance of importance in a few cases - a clear distinction from crop species, where disease and insect resistance predominate as selection criteria. The poplars are exceptional industrial species in many respects:- hybridisation and clonal selection in much narrower populations has been the traditional approach to breeding, and, perhaps not coincidentally, disease resistance as a major selection criterion. Significant genetic gains are being achieved in these breeding programmes but, in particular for the long rotation species, there has been only a minor impact to date on the genetic quality of plantations. Major limitations to rapid improvement with most of the established industrial species are:

- Long generation intervals, related to poor juvenile-mature correlations and the long juvenile phase with respect to flowering.
- Low effectiveness of selection for many characters, due to low heritability or difficulty in assessment.
- Through the use of open-pollinated orchards, the exploitation of only a part of the genetic variation available.

Major research priorities for these established industrial species should be the broader development of methods for the propagation of full-sib families and clones, and the development of methods for early and more accurate selection and the promotion of precocious flowering. While only a small proportion of current industrial plantations are located in developing countries, it is clear that much of the additional area of up to 100 million hectares required worldwide will be situated in tropical areas. Established, proven industrial species are likely to comprise a major proportion of this additional estate. A major practical tree improvement objective will be the establishment of sound breeding programmes for these new plantation operations.

In order to make use of a broader range of sites, and to supply products currently provided by harvesting in natural forests, it is likely that a significant proportion of the additional 100 million hectares will be established with tropical species which are not widely used or known at present. Selection criteria for these "new" industrial species will be similar to those for established species. Some are probably very amenable to improvement, while others present problems, e.g. flowering and seed problems, and insect susceptibility. Many potentially valuable species are simply not well known, in terms of their adaptation and biological and genetic features, and gene pools are probably under threat. The broader implementation of good tree improvement programmes will be an important priority for these species. The testing of potentially useful species, characterization of mating systems, provenance collection, establishment, maintenance and assessment of trials, implementation of gene conservation measures, and commencement of other breeding work, poses a very large task.

There is a global requirement also for several hundred million hectares at least of non-industrial plantings, mainly located in developing countries. Breeding programmes for non-industrial trees differ markedly from those for industrial species in the wide variety, and the complexity, of selection criteria. Several of the taxa are highly variable and display features which render them particularly amenable to rapid improvement by traditional means. Biological features of many potentially valuable non-industrial species, however, remain largely unknown. Although not well studied, some gene pools are under threat, and some species display unpredictable patterns of variation as a result of human disturbance. Although selection work has been undertaken in a few programmes, most non-industrial species remain at the species testing stage. The need to establish plantings on a wide range of sites, many marginal or difficult, dictates that non-industrial plantings are likely to be reliant on a substantial number of the thousands of potentially useful species. The work involved in selecting the most promising species for each site is formidable. Major emphases of improvement for non-industrial species are likely to be taxonomic studies of variation, species and provenance testing, the assessment of reproductive features, and conservation activities. Interspecific hybridization is likely to be of value for several taxa. Where regionally consistent selection criteria can be defined and the tree crop is considered sufficiently important, clonal selection programmes are likely to be of value for species easily propagated vegetatively, and seed tree selection for species propagated by seed. Sophisticated breeding programmes are unlikely to be warranted for most species. It is possible that a "once only" or "quantum leap" approach to breeding may be more appropriate than breeding programmes offering continuous incremental gains.

As pointed out by Kanowski (1993), tree improvement is greatly under-resourced. This is especially so in developing countries, where funding provided through national, regional, bilateral and international programmes is not sufficient to conduct properly the essential traditional tree improvement activities outlined above. The problems are further compounded by inadequate levels of training and facilities in most areas. For the non-industrial species in particular, it is these resource limitations and the variability in user requirements, rather than biological constraints, which constitute the major impediment to rapid improvement.

BIOTECHNOLOGIES AND THEIR APPLICATIONS TO TREE IMPROVEMENT

Cryopreservation and in Vitro Storage

For material displaying regenerative competence (e.g. embryogenic cultures), a research project directed towards the development of procedures for successful *in vitro* storage or cryopreservation of cultures for a new species might be expected to be short term (say three to five years), with a moderately high expectation of success.

Although increasingly used operationally for the storage of threatened germplasm of agricultural species, *in vitro* storage and cryopreservation have little to offer for this purpose in forest tree species. Gene pools of most of the established industrial species are reasonably well conserved in *in situ* and *ex situ* stands and preserved in seed stores. Undoubtedly, an urgent gene conservation problem exists for many tree species, particularly among the tropical hardwoods and non-industrial species. Distributions of these species are poorly known, as are their biological characteristics. Although seeds of some species will be recalcitrant, many are orthodox. Major impediments to the conservation of forest tree germplasm are the availability of resources sufficient for only a very small fraction of the survey and collection work which would be required before any germplasm could be stored, and the unreliability of many existing seed storage facilities. Replacement of existing facilities with more sophisticated technology is not likely to make a positive impact on the gene conservation problem with tropical tree species. Even for the recalcitrant species, activities should favour the establishment of *ex situ* plantings which will facilitate evaluation of the material. In the longer term, cryopreservation and *in vitro* storage may have some application as a back-up conservation strategy for populations of well surveyed species which are recalcitrant. The poplars and walnut currently fall into this category. As has been pointed out by Wang *et al.* (1993), *ex situ* conservation measures should themselves be regarded only as complementary to sound *in situ* programmes.

Cryopreservation warrants much more attention as a means of maintaining juvenility and capturing genetic gains offered by clonal forestry with industrial species. The technology is thus applicable mainly to programmes where good breeding programmes are in place, clonal forestry is a realistic goal, and "rejuvenation" is difficult - in particular for the conifers.

With the increasing emphasis on international breeding of both industrial and non-industrial forest tree species in tropical countries, international movement of material will be important. In the short term, such movement is likely to be mainly in the form of seed for species and provenance testing. Movement of vegetative material, however, is likely to become more important as breeding programmes advance, and *in vitro* approaches could find application in the longer term.

Molecular Markers

Techniques for the analysis of isozymes, RFLPs and RAPDs are now well established, and the adaptation of procedures to suit new species could be expected to be a relatively short term undertaking. Markers have important immediate applications in supportive research for advanced breeding programmes with industrial species - mainly in relation to quality control, e.g. checking of clonal identification, orchard contamination and within-orchard mating patterns by "fingerprinting". Isozymes will be satisfactory for many, but not all, of these purposes.

Markers also have important immediate application in supportive research for tropical hardwoods and non-industrial species, in particular for essential taxonomic studies and investigations of mating systems. Both isozymes and DNA markers will be useful for these purposes. Markers will also be useful for the quantification of genetic variation to aid in sampling strategies for gene conservation and breeding population collections, although, due to modest correlations with patterns of variation for adaptive traits, they must be used conservatively as complements to traditional methods.

Realistically, application of marker-assisted selection in the short or medium term is likely to be very limited. Cheaper markers would be required and, even when these were available, the technology would apply mainly to advanced and sophisticated breeding programmes - those in which creation and maintenance of the appropriate population structures could be afforded, and where clonal forestry is achievable. A small number of programmes fall into this category and, for these, some effort aimed at developing marker-assisted selection can be justified. For most species, current resources would be far better directed towards moving breeding programmes to this stage of advancement, rather than to the development of marker assisted selection. Marker assisted selection is unlikely to have much application to non-industrial species, although, by virtue of a very short generation time, taxa such as *Gliricidia* may be useful model species for experimentation.

The major current value of markers lies in long term strategic research - in the great contributions which marker studies are making to advances in the understanding of basic genetic mechanisms and genome organization at the molecular level. For forest tree species, the study of quantitative traits will be an important emphasis of this work in coming years. This work will be most efficiently concentrated on a few model species, e.g. *Pinus taeda* and in particular hybrids such as *P. elliottii* × *P. caribaea*.

In Vitro Selection

Many recent publications concerning crop plants have reported useful correlations between *in vitro* responses and the expression of desirable field traits, most commonly disease resistance, although positive results are available also for tolerance to herbicides, metals, salt and low temperatures. For the selection criteria of major general importance in forest tree species, in particular vigour, stem form and wood quality, however, poor correlations with field responses will limit the usefulness of *in vitro* selection. *In vitro* selection is therefore likely to have very limited application in forest tree species. It will be of possible interest in a few programmes where there is a disease resistance selection problem, but of no broad strategic value as a research objective.

Genetic Engineering

Crops transformed with genes for insect and virus resistance and resistance to various types of herbicides are at or near commercial application, and plants into which these genes have been inserted include the poplars. Many projects are in progress with forest tree species, in particular for modification of lignin biosynthesis through antisense technology. Insertion of the insect or herbicide resistance genes currently available into a new species would constitute a major research undertaking, and successful application would be dependent on being able to regenerate from the transformed cells. Manipulation of more complex traits would be a much more formidible undertaking, and, although rapid progress is being made in this field, much research remains to be done. The availability of effective transformation techniques remains an obstacle, but improved techniques are being developed.

Regeneration is a difficulty for some tree species, but the problem may be over-rated - the non-competence of mature material is not necessarily an obstacle to effective application of genetic engineering, provided that juvenile material responds satisfactorily. An often overlooked research component is the testing which would be required before a responsible recommendation for large scale deployment of trangenic plants could be made. Such testing could be extensive and prolonged, depending on the species and genes involved. Research projects of this type are necessarily intensive, and must be regarded as long term with only a modest expectation of success. Insect resistance is of potential value, e.g. in the poplars and some tropical hardwoods. A single gene may be sufficient to confer resistance for very short rotation species, but a much more cautious approach should be applied with long rotation species. The work involved in introducing several different resistance genes, sufficient to ensure that insects do not acquire tolerance during a long rotation, should not be underestimated. The reduction of lignin biosynthesis is a very valuable objective for the pulp species. The introduction of herbicide tolerance genes is of some interest, but in many programmes the advantage of using unguarded herbicide application may not be sufficient to pay for the research program. The extent to which genetic engineering permits substitution of environmentally "friendly" herbicides such as glyphosate for currently used residual herbicides could be an important factor.

Cold tolerance genes are likely to be of some commercial value in many species, in particular the eucalypts. Much remains to be done, though, to establish that sufficient tolerance can be conferred using antifreeze proteins, and to extend the work to tree species. Prevention of the escape of genes into wild populations is likely to become an important concern, and sterility should be an early target of genetic engineering work with forest tree species. The major factor limiting application of genetic engineering in forest tree species is the state of knowledge of molecular control of the traits which are of most interest - those relating to growth, adaptation and stem and wood quality. Genetic engineering of these traits remains a distant prospect.

It is important that genetically engineered genotypes be of high quality with respect to other traits as well. The clonal test is the most logical basis for integration of genetic engineering into traditional tree improvement programmes. For these reasons, genetic engineering is most appropriately conducted with species where breeding programmes are advanced and clonal forestry can be realistically contemplated. Research on this subject should not assume a high priority with species for which natural variation available within the taxon remains poorly investigated.

Somaclonal Variation

Variation induced during cell or callus cultures has been reported for many species. For some crops, variants have been produced showing economically useful levels of traits such as resistance to disease and increased levels of salt. The phenomenon is still not clearly understood, and persistance of modifications through subsequent sexual generations has been demonstrated in some cases but not in others. This is not a research area where favourable results could be predicted with any confidence, and use of the approach is furthermore dependent on being able to regenerate whole plants from the cells or callus. Research with new species will be more soundly based when the phenomena are better understood in the model species already under study. Assuming stability of the characteristics, the approach is of most value where selection can be done at the cell level (e.g. by exposure to a phytotoxin or to high mineral levels), thus enabling screening of large numbers of genotypes, and where the level of the trait sought lies outside the range available naturally in the species. Cold tolerance in the eucalypts is an example which may meet these criteria. No immediate applicability is evident to the little known tropical hardwoods or to non-industrial species, for which genetic variation naturally available is generally poorly defined.

Protoplast Fusion

This is a field with a long history of research. There have been several successes, notably with species of the Brassicaceae and Solanaceae. Severe limitations are imposed, however, by the taxonomic relationship of parents, and the requirement for regeneration from protoplasts. Expectations of success with a new pair of species could be expected to be low. This is a field with little apparent applicability to forest tree species, particularly those (e.g. some tropical hardwoods and non-industrials) for which investigations of the possibilities for traditional hybridization remain minimal. In the longer term, the objectives of protoplast fusion may be better served by manipulations at the DNA level.

Gametophyte Cultures

Although anther culture has been used for the rapid production of homozygous lines in the breeding of some self-pollinated cereals and vegetable crops, few reports exist of regeneration from gametophyte cultures of forest tree species. Research with this objective would be likely to be long-term with a relatively high risk of an unsuccessful result. Induction of haploid plants has no immediate application in forest tree improvement programmes. Such plants may be of some use in basic genetic studies, e.g. for studies of heterosis in forest tree species. As a long term strategic research objective for industrial species, induction of haploid plants should be of low priority until such time as methods for early selection and the promotion of early flowering are available. The breeding of non-industrial species is unlikely to warrant this level of sophistication.

In Vitro Embryo Rescue

This technique has been used, particularly in fruit trees, to grow embryos that normally would abort due to incompatibility between ovule and embryo development, and also for the rescue of the zygotic embryos of apomictic species. The techniques are not difficult, and development of protocols for a new species generally would be a minor research task - short term with a high expectation of success. Embryo rescue has been used occasionally in forest tree species, but the requirement for such technology is likely to be limited - to what probably will be a small number of hybrids which are sufficiently close to produce a normal embryo but for which embryo development *in vivo* is restricted. In the short term, research of this type is likely to have a very low priority. In the longer term, as barriers to natural hybridization become better defined, some work can be targeted at hybrids which species testing has identified as potentially of interest and for which other studies suggest that *in vitro* approaches might provide a solution.

Micropropagation

Over 1 000 plant species have been micropropagated, including over 100 forest tree species. It is reasonable to conclude that, with sufficient research effort, successful protocols could be developed for most tree species. In general then, a research project of this type might thus be described as short to medium term with a moderately high expectation of success.

For most industrial species, costs of planting stock and insufficient data regarding field performance remain major obstacles to be overcome before broader use of micropropagules as direct planting stock could be contemplated. Micropropagation has an immediate application in integrated clonal propagation systems featuring the commercial planting of cuttings harvested from rapidly multiplied, micropropagated stool plants of the selected clones. This approach is of value only in very advanced breeding programmes incorporating the identification of outstanding clones - currently only a few programmes with tropical eucalypts, and perhaps also some poplar programmes, but potentially also with other industrial species.

Appropriate integration into breeding programmes is essential. Where clonal testing on a reasonable scale is possible and affordable, the current applicability of protocols mainly to juvenile material is not necessarily an impediment to the capture of good gains through clonal forestry. This conclusion, however, is dependent on the ability to store juvenile material for the period of a clonal test. Genetic variation in response, often substantial, is not likely to be a major problem where clonal testing can be preceded by screening for responsive genotypes, although demonstration of the absence of adverse correlations with economic traits is important. Breeding programmes with new industrial species and non-industrials are not sufficiently advanced to warrant much use of micropropagation in the short term.

Micropropagation may have wider application for the multiplication of stool plants of industrial species as breeding programmes become more advanced and other limitations to clonal forestry (e.g. maturation problems) are overcome. For some non-industrial tree species, micropropagation may ultimately have a role in the multiplication of selected varieties prior to release. Development of simple micropropagation protocols for those species for which such are not already available is therefore a useful research objective, but one which should not take priority over issues such as advancement of the breeding program. With considerable research, developments with somatic embryogenesis and artificial seed technology may overcome the planting stock cost limitation discussed above, and enable direct use of such propagules in plantation establishment. For industrial species then, development of these technologies is a useful long term research objective, but one which is best pursued with one or two appropriate model species, e.g. *Picea abies* and *Pinus taeda*.

IN VITRO CONTROL OF THE MATURATION STATE

A long history of research with *in vitro* rejuvenation has led to some successes, but little evidence that rejuvenation can be completely, permanently and reliably achieved by this method. Similarly, sporadic reports of accelerated maturation through *in vitro* manipulations are not encouraging with respect to reliable acceleration of ageing. Further empirical work with these objectives is likely to have a low probability of success. An understanding of the molecular basis of maturation is much more likely to lead to practical manipulation, but this work is in its infancy, and acceleration or reversal of maturation to precise levels remains a distant prospect.

For clonal forestry with industrial species, maintenance of juvenility as is about as useful as rejuvenation for many purposes, and probably achievable using technologies such as cryopreservation or stimulation of juvenile stump sprouts. Nevertheless, more fundamental control of the maturation state remains one of the most valuable objectives of long-term strategic research in forest tree improvement with industrial species. Rejuvenation is most applicable to programmes where good breeding programmes are in place, and where other limitations to clonal forestry do not exist. Maturation is a much less important issue with many of the non-industrials. Manipulation of the maturation state to induce early flowering and reduce generation intervals is potentially of greater interest than rejuvenation, for industrial species at least, but only of real value where active breeding programmes are in place.

SYNTHESIS

Biotechnology represents, for many agricultural crops, the best hope for meeting the urgent objectives of breeding programmes - conservation of gene pools of wild relatives, multiplication of elite cultivars, and, in particular, the acquisition of virus and insect resistance. This applies also for many crops in developing countries, including cassava, a crop grown almost entirely by resource-poor farmers. By contrast, biotechnology offers little in relation to the most urgent priorities in forest tree improvement, although forest tree species are the subject of very active biotechnology research programmes. This is because of the major objectives of most tree improvement programmes differ greatly from those for crop species, because of the long generation times of trees, because of the great number of tree species which can provide a given end use product, because of the relatively low value of the product, and a number of other factors.

Possibilities for technology replacement or supplementation in tree improvement programmes in the short term are:

- *For breeding programmes with established industrial species:*
  - The use of molecular markers in quality control in advanced breeding programmes, e.g. for checking of clonal identification, orchard contamination and within-orchard mating patterns by "fingerprinting".
  - The use of micropropagation in integrated clonal propagation systems featuring the commercial planting of cuttings harvested from rapidly multiplied, micropropagated stool plants of the selected clones. This approach is of value only in very advanced breeding programmes incorporating the identification of high quality clones.
- *For breeding programmes with "new" industrial and non-industrial species:*
  - The use of markers in essential taxonomic studies and investigations of mating systems.
  - The use of markers for the quantification of genetic variation to aid in the design of sampling strategies for gene conservation and breeding population collections.

Strategic research priorities relating to the application of biotechnology in tree improvement are:

- *Long-term generic research. Priority objectives are:*
  - Genetic engineering for sterility. This is of high priority because it will underlie many of the eventual applications of genetic engineering.
  - The use of molecular markers and DNA transformation techniques to investigate genetic processes at the molecular level, in particular those relating to complex traits such as growth, adaptation and stem and wood quality. Of high priority, this work is of relevance in particular for industrial species, but will pave the way also for applications of biotechnology to non-industrial trees.
  - Molecular studies of the maturation state. This is of high priority for industrial plantation species.
  - Development of somatic embryogenesis, in combination with artificial seed technology, as a cheap clonal propagation method for industrial plantation species. This is of moderate priority.

Research projects of the above type are most efficiently conducted with a small number of model species. Diffusion of resources and effort to many species is likely to impede progress.

- *Long-term specific research. Some objectives are:*
  - Genetic engineering of useful traits, e.g.:
  - lignin reduction in pulp species
  - cold tolerance, particularly in eucalypts
  - insect resistance, e.g. in poplars and perhaps Meliaceae (when appropriate breeding programmes are in place)

Transformation with appropriate genes may be achieved within the short to medium term (say the next five to ten years) but must be followed by perhaps ten years of field testing before responsible commercial deployment could be recommended.

- Marker-assisted selection, for species where breeding is advanced and where creation and maintenance of the appropriate population structures is feasible and affordable. It will probably be several years before this is possible on an effective operational scale.
  - *Short- to medium-term research. Useful objectives are:*
- The examination of genetic correlations between regenerative competence and commercially important field traits. This is of high priority.
- The development of cryopreservation methods as a means of maintaining juvenility, for advanced breeding programmes with industrial species.
- The development of cryopreservation as a backup measure for gene conservation in proven species for which breeding programmes are in existence and for which seed recalcitrance has been demonstrated (moderate priority).
- The development of simple micropropagation techniques for species for which such is not already available (low to medium priority).

There are currently few areas where biotechnology can profitably replace existing technologies in tree improvement. The potential applications of several biotechnologies are interdependent e.g. the application of genetic engineering, and of cryopreservation for the retention of juvenility, are dependent on the availability of a suitable method for clonal propagation from cultured material. The applications of these technologies, and marker-assisted selection, are really only appropriate to operational, sound clonal forestry programmes, and it can be argued that clonal forestry is the gateway to major applications of known biotechnologies in forest tree improvement. Sound clonal forestry is dependent on the existence of good breeding programmes to produce genotypes for their use in well-planned and sound tree planting programmes. The funding of biotechnological research initiatives cannot be at the expense of the development of good genetic improvement programmes. Apart from the large investments required for the development of most biotechnologies, and for appropriate testing, large investments in many tree breeding programmes are required to reach the appropriate levels of advancement, frequently with added levels of structural sophistication. As emphasized by Burdon (1992) then, a commitment to biotechnology must be part of a substantial increase in the total commitment to genetic improvement. This level of investment in breeding is not being made for most industrial species, and may never be affordable for most non-industrials. The gap between traditional methods and biotechnological approaches to forest tree breeding is therefore substantially a matter of financial investment.

The high costs of the required research programmes argue strongly for collaboration, rather than competitive proprietary biotechnology, as stressed by Burdon (1992). Certainly, it is desirable at least that expensive and long-term generic research be conducted collaboratively by specialized laboratories and with appropriate model species.

BIOTECHNOLOGY AND TREE IMPROVEMENT IN DEVELOPING COUNTRIES

Many highly profitable private forest plantation programmes exist in developing countries. Taking into account risks and potential commercial benefits, managers of these programmes will make their own decisions on the appropriate stage at which investment in biotechnology is appropriate, and the level of investment. These industrial operations aside, funding available for genetic improvement in developing countries, in particular for "new" industrial and non-industrial species, is grossly inadequate for the urgent exploration, conservation and testing work which needs to be done, and for the training and facilities which are required. Funding currently used for these purposes should be redirected to biotechnology only to the extent that effective technology replacement is achievable in the short term. In this category are the use of molecular markers in studies of reproductive biology, in taxonomic investigations, and for the quantification of genetic variation. Massive increases in funding for tree improvement in developing countries may warrant more attention to biotechnology, but diversion of currently available funds to longer term biotechnological research initiatives is likely to have a negative impact, overall, on genetic improvement programmes for these species.

CONVENTIONAL BREEDING

Conventional plant breeding involves crossing carefully chosen parent plants, then selecting the best plants from the resulting offspring to be grown on for further selection. For cereals, hundreds of individual crosses are carried out by hand to create seed for the first filial (or F1) generation. The resulting F1 plants are uniform, but in the following generation several hundred thousand different plants are produced. Because of the way genes work, the new combinations produced from each cross are not revealed until the second (F2) generation. It is this enormous diversity of new gene combinations that may hold the key to a successful new variety. The plant breeder's task is to select the plants most likely to meet his breeding objectives. Seed from the best of these F2 plants is grown on in small rows or plots and the best plants again selected – this process is repeated year after year until only the very best plants remain. As promising new lines emerge, tests are conducted on each plot to assess factors such as yield, disease resistance and endues quality. Once the best lines are purified to ensure that every plant has the same characteristics, the process of multiplying seed begins. These 'inbred' lines are then ready for entry into official trials, some six to ten years after the initial cross.

Hybrid Breeding

For some crop species, the seed supplied to growers is that produced from the first cross between selected parents. The resulting varieties, known as F1 hybrids, offer potential advantages in crop performance. Hybrid breeding is widely used to produce varieties of field vegetables, maize and oilseed rape. F1 hybrids are unique in expressing 'hybrid vigour' in the growing crops for a single year. This may result in higher yields, greater uniformity, or improvements in quality. Unlike inbred lines, F1 hybrids do not breed true year after year and their performance gains are not maintained in subsequent generations.

Improved Breeding

With increasing knowledge and improved technology, breeders have developed ways to enhance the speed, accuracy and scope of the breeding process. There are several ways to reduce the lengthy interval between the first cross of selected parents and establishing true breeding lines of promising new varieties. For example, maintaining parallel selection programmes in northern and southern hemispheres allows two generations to be produced each year. Single seed descent produces very small plants under restricted growth conditions. Large numbers of these plants are cultivated in artificial growth rooms, with two or more generations produced in a year. In potatoes, mini-tuber breeding speeds up the slow multiplication process by producing miniature plants under greenhouse conditions. More recent laboratory techniques enable breeders to operate at the level of individual cells and their chromosomes. In certain crop species, such as potatoes and oilseed rape, it is possible to produce new varieties in the laboratory through

Protoplast Fusion

Individual plant cells with their outer walls removed are fused, and the fused cells then induced to divide and grow in a culture medium. Whole plants are eventually regenerated containing new combinations of genes from the two parents.

EMBRYO RESCUE AND ASSISTED POLLINATION

Allow breeders to expand the range of available characters by making crosses between species, which would not produce viable offspring outside the laboratory.

Double Haploid Breeding

Enables breeders to produce genetically uniform lines within one generation. This effectively by-passes the lengthy process of self pollination and selection normally required to produce true breeding plants. Latest developments in genetic science have greatly improved our understanding of how plants behave, offering additional ways to enhance the breeding process.

Genomics

By mapping the genetic makeup, or genome, of crop species, scientists can identify the exact position and function of individual genes. Genome mapping has revealed striking similarities in the genomes of different crop species, such as rice, wheat, barley and rye. This information is already helping to broaden the scope and precision of current breeding programmes.

Marker Assisted Breeding

Allows breeders to determine whether desired traits are present in a new variety at an early stage in the breeding programme.

Genetic Modification

Genetic modification of plants is achieved by adding a specific gene or genes to a plant, or by knocking out a gene with RNAi, to produce a desirable phenotype. The resulting plants are often referred to as transgenic plants. Genetic modification can produce a plant with the desired trait or traits faster than classical breeding because the majority of the plant's genome is not altered.

To genetically modify a plant, a genetic construct must be designed so that the gene to be added or knocked-out will be expressed by the plant. To do this, a promoter to drive transcription and a termination sequence to stop transcription of the new gene, and the gene of genes of interest must be introduced to the plant. A marker for the selection of transformed plants is also included. In the laboratory, antibiotic resistance is a commonly used marker: plants that have been successfully transformed will grow on media containing antibiotics; plants that have not been transformed will die.

In some instances markers for selection are removed by backcrossing with the parent plant prior to commercial release. The construct can be inserted in the plant genome by genetic recombination using the bacteria Agrobacterium tumefaciens or *A. rhizogenes*, or by direct methods like the gene gun or microinjection. Using plant viruses to insert genetic constructs into plants is also a possibility, but the technique is limited by the host range of the virus. For example, Cauliflower mosaic virus (CaMV) only infects cauliflower and related species. Another limitation of viral vectors is that the virus is not usually passed on the progeny, so every plant has to be inoculated. The majority of commercially released transgenic plants, are currently limited to plants that have introduced resistance to insect pests and herbicides. Insect resistance is achieved through incorporation of a gene from Bacillus thuringiensis (Bt) that encodes a protein that is toxic to some insects. For example, if cotton pest the cotton bollworm feeds on Bt cotton it will ingest the toxin and die.

Herbicides usually work by binding to certain plant enzymes and inhibiting their action. The enzymes that the herbicide inhibits are known as the herbicides *target site*. Herbicide resistance can be engineered into crops by expressing a version of *target site* protein that is not inhibited by the herbicide. This is the method used to produce glyphosate resistant crop plants.Genetic modification of plants that can produce pharmaceuticals (and industrial chemicals), sometimes called *pharmacrops*, is a rather radical new area of plant breeding

Proteomics

Allows breeders to understand how genes behave in different parts of the plant and under different growing conditions.

Maintaining Genetic Diversity

Maintaining biodiversity is central to the process of crop improvement. It is in every breeder's interest to ensure that the gene pool from which new traits are selected remains as extensive as possible. Plant breeders created the first gene banks in the 1930s to conserve the valuable genetic diversity within past and present varieties, as well as landraces and wild relatives of cultivated crop species. Plant breeding is integral to ongoing initiatives to identify, classify and conserve existing biodiversity.

Testing Plant Varieties

Before any new crop variety can be placed on the market, it must undergo statutory testing under a process known as National Listing. Successful varieties are placed on a National List or register of varieties approved for marketing. National Listing rules are determined on a European basis, and apply to all the major agricultural and vegetable crop species. Official trials are conducted, in most cases for a minimum of two years, to test each new variety for a range of characteristics which together determine its uniqueness, its genetic uniformity, and its value to growers and the rest of the food chain.

National Listing is extremely rigourous – the majority of varieties entered do not complete the process. In winter wheat, for example, only a quarter of varieties entered for National Listing during the 1990s were finally approved.

National Listing – DUS

All varieties entered for National Listing are assessed for Distinctness, Uniformity and Stability (DUS). In the case of cereals, some 30 individual characteristics of the plant are inspected to verify that it is distinct, ie clearly distinguishable from other varieties, that its characteristics are uniform from one plant to another, and that the variety is stable in that it breeds true to type from one generation to the next.

National Listing – VCU

For agricultural crops, National Listing also involves trials to establish a variety's Value for Cultivation and Use (VCU). This provides an assurance to growers that only varieties with improved performance or end-use quality can be officially approved for sale.

Recommended and Descriptive Lists

Once a new variety has been added to the National List, it is cleared for marketing. However, its success in the market place is by no means guaranteed. Further independent trials are conducted each year to compare the performance and quality of the best varieties.

These trials provide the basis for detailed information and advice to growers and their customers. As shown in the following illustration for winter wheat, the process of statutory and commercial evaluation can take up to five years. Only a handful of varieties clear all these hurdles, which are in addition to the many years spent testing and selecting in the breeder's own trials programme.

No GM crops can be marketed until they have been assessed and approved in terms of human health, food safety and the environment. The cross-border movement of genetically modified organisms (GMOs) is regulated at a global level under the internationally agreed Biosafety Protocol.

Variety Maintenance and Identity Preservation

As well as developing new varieties, plant breeders maintain the genetic purity of existing lines and pre-commercial seed supplies year by year. This process is costly and time-consuming, but essential to maintain the quality and performance of each variety. For cereals, variety maintenance begins after a few years of selection trials, when the promise of a variety is just emerging. At that stage, all that exists of what may become a widely grown variety is a single row containing around 100 plants.

The breeder then bulks up supplies of the purified lines of breeder's seed into prebasic and then basic seed. Each year specialist seed growers are used to grow basic seed for the first generation of certified seed – or C1 seed. After one more year this becomes C2 seed, the main source of certified seed used by farmers. The plant breeder continuously maintains breeder's seed for the process of multiplication through pre-basic, basic, C1 and C2 seed to ensure the variety's performance and quality year after year. Greater emphasis is now being placed on preserving the identity of individual varieties after harvest, both to conserve quality characteristics and to meet consumer demands for assurances about the integrity and traceability of their food.

Seed of an approved variety can only be marketed if it meets strict quality criteria Seed quality standards are laid down in UK and EU law, and policed by agencies appointed by Government. The UK's official seed certification system offers an independent assurance of quality to growers. Minimum standards apply for varietal identity, purity and germination capacity. In addition, strict limits apply to seed-borne diseases, and the presence of physical impurities such as weed seeds. Around 9 per cent of the UK arable area is used to multiply the pure lines of seed from the plant breeder into certified seed. Several thousand individual crops are involved; each grown under specific management regimes to ensure the purity and integrity of the resulting seed is maintained. To gain certification, every seed crop must undergo crop inspection and seed testing. Seed certification underpins the health and purity status of the major arable crops in Britain. It offers an independent benchmark of quality on which buyers of seed and their customers depend.

For certain crop species – particularly small-grain cereals – growers can opt to save their own seed for sowing the following year provided care is taken to ensure that the crop remains healthy and free from impurities, and that the resulting seed is carefully conditioned and cleaned. Without independent testing for germination and freedom from seed borne diseases, however, farm-saved seed can harbours risks to growers that may not become apparent until well into the growing season. Many growers choose certified seed for the peace of mind that quality and performance is independently assured.

POLYPLOID BREEDING

Cytological study of polyploid species shows why most of them behave in bteeding like the hybrid polyploid of the Primula kewensis type. Whether they are tetraploids, hexaploids or octoploids, their chromosomes usually unite in pairs at reduction. This rule is particularly strict in those plants which depend on seed fertility for their reproduction, e.g., the cereals.

In these the differentiation is sharper even than in Primula kewensis. It is more like that in the Raphanus-Brassica tetraploid, where no pairing takes place between corresponding chromosomes of the two species even in the diploid hybrid. They are, as it were, in two watertight compartments. The species behave on self-fertilisation almost exactly as though they were diploids. Yet, apart from their chromosome numbers, there are three ways in which these species show us that they are polyploids. In the first place, a particular character in a polyploid may be given by the presence of any one of several different factors. These factors are probably in corresponding chromosomes of the different diploid ancestors of the polyploid. In the hexaploid bread wheat, Tritkum vulgare, Nilsson-Ehle found that some varieties may have three independent factors affecting the red colour of the grain.

In the hexaploid oats, Akerman found similar factors affecting the presence of green leaf-colour. Other characters are affected by-two factors. We do not know that these multiple factors are due to the plant being a polyploid, but since they are not usually found in diploid plants, it is very probable. The occurrence of two independent factors for black grain-colour has been turned to advantage in breeding oats in Sweden. Old black varieties of oats had a single factor which may be called BXBX for black colour.

These varieties used to sport so as to give a hundred or so white grains in every bushel, owing to the corresponding chromosomes in the second set carrying the recessive factor b9 and occasional pairing taldng place between the B and b chromosomes. The white grains, of course, reduced the value of the oats by giving it unfairly an appearance of impurity. Akerman x therefore crossed two black varieties, and from the F2 a new type doubly black BXBXB2B2 was raised, which never sports to white seeds.

Secondly, when diploid species are X-rayed their offspring change in every direction; polyploids do not vary anything like so much as their diploid relatives. While tetraploid species of oats and wheat did mutate a little under the same doses as gave big changes in the diploids, the hexaploid bread wheat and cultivated oats showed no change whatever. What is true of change induced by X-rays is probably equally true of natural change.

The extra chromosome sets act as a buffer against the effects of changes in the chromosomes, so that in polyploids we can have very little idea of what is going on inside the chromosomes, as compared with diploids. The doubly black-seeded variety of oats is an example of this; the second pair of factors is there for safety and if mutation of Bx to bx occurs, we shall not notice it.

But we can occasionally see the results of other changes, which provide a third kind of evidence of polyploidy. Diploid pure lines sometimes breed true, and for generations show no variation to the closest scrutiny: such are known in the Triticum monococcum, of de Vilmorin, which has bred true for eighty years, and in Johanssen's pure lines of broad bean, Vicia Faba. But the hexaploid cereals cannot be depended upon. Awnless cultivated oats of as pure race as can be got always throw a proportion—varying with the variety and particularly high in new varieties—of individuals having awns on their flowering glumes.

And when these are selfed, their progeny (which will appear in cultivation unless oats for seed are severely rogued) are of quite a distinct and inferior type, indistinguishable, in fact, from the wild oat, Avena fatua. These are called "fatuoid" oats. They are characterised by having both the spikelets provided with a twisted, kneed awn, and each grain marked by a prominent articulation surrounded by a tuft of hairs. The chromosome behaviour at reduction that the "mutants" are due to the same cause as the appearance of the mealy offspring of P. kswensis, which also resemble a related species, P. verticillata* In both species of Avena, A. fatua and A. sativa each chromosome type is represented by three pairs. Taking two of these pairs -we may consider that A. fatua has the constitution FF9FF9 while A.

Sativa has the constitution FF9ff9 and regularly produces gametes Ff which reproduce it truly. Both, as a rule, breed true, but occasionally in A. sativa (particularly in new varieties which have been disturbed by hybridisation) F pairs with/, or the four chromosomes FFff form a ring at reduction. The result is the formation of germ-cells that are FF or ff instead of the normal Ff.

When these meet the normal germ-cell the result is the production of two new types—FFFf and Ffff. The first of these is our slightly awned type, which when selfed gives, together with the normal and its own type, the pure fatuoid, FFFF, in the proportions 1:2:1. The same result is given by the first cross between the wild oat and cultivated oat, so that we know in this case roughly what makes the difference between the two.

This property of giving away its hybrid origin is probably very common in new polyploid species, for the bread wheats do just the same thing. They throw a sport called a "speltoid" in exactly the same way as the oats throw a fatuoid. Again, this form resembles a related species, Triticum Spelta, in the character of its spike and grain. Such behaviour shows that rogueing is a much more serious business with polyploids than with diploids. All the materials are there for them to produce a whole array of sports, and if they are left to themselves although they are continually self-fertilised they will none the less gradually become more and more adulterated by their own throw-backs, which will mainly be of a character unwanted in cultivation.

CROSSING POLYPLOID SPECIES

If this is the way polyploid species behave when they ate continually inbred, something rather complicated is to be expected when they are crossed. Diploid species-crosses very often give sterile hybrids. Polyploid species, if they have the same number of chromosomes, cross more often to give fertile hybrids. Polyploid species are therefore often more difficult to separate than diploid. Since it is not known how the polyploid species themselves are made up (except that they are hybrid), it cannot be predicted how they will behave when they are bred together; the same combination may be broken up in many different ways.

One of the exceptional results they give is what Engle-dow has called "shift." In the offspring of a cross between Triticum polonicum and T. durum he found no segregation of the long-glume character of the Polish wheat. All the progeny for several generations remained intermediate in character. In the same way, in a cross between a rough-chaffed rivet wheat, T. turgidum, and T. polonicum, Biffen never recovered the grey chaff colour on the rivet, although progeny were grown to the number of 100,000. The same suppression of a character by crossing has been found by Backhouse and Vavilov in the tetraploid cereals, and in other cases they have found the reverse effect of a character being accentuated.

Take the case of the cultivated strawberry, whose history have been well described by Bunyard and Pearl. The wood strawberry, Fragaria vesca, was cultivated for its fruits in the fifteenth century. It produced no variations of importance;, except an ever-flowering "alpine strawberry" analogous to the "All Saints" cherry and the "Lloyd George" raspberry. The fruits always remained small. This species is diploid. Another species that came into cultivation at this time in France and England was Fragaria elatior, the Hautbois strawberry, also a native of Europe. This species differs from the former in holding its ripe fruits above the leaves, and in the musky flavour of the fruits. Like the wood strawberry, it has small fruits, and although still occasionally grown, it likewise has shown no considerable improvement in size in the course of four hundred years of cultivation.

This species is usually dioecious, having male and female flowers on different plants. F. elatior is hexaploid, and therefore will not yield fertile hybrids with F. vesca. In the seventeenth century, the scarlet strawberry, F.virginiana, from North America was introduced by John Tradescant. The plant was distinct in vegetative characters and its fruit was different in shape and flavour from that of the two European species, but it was scarcely any bigger, about half-an-inch long. It produced more varieties than the older strawberries, but no marked improvement. This species is octoploid and does not cross with either of the first two.

The last important species to be introduced to Europe was the South American F. chiloensis which came to France in 1712. But the plants introduced were all females, and consequently set no seeds until planted near other species with pollen. F. chiloensis—very distinct from the others, having stout scapes with few flowers, and fruits twice their size—is octoploid. From all accounts it is certain that the few plants of it that were introduced afforded the first possibility of raising the large-sized fruits familiar in our modern cultivated varieties. But since the plants were female, it is evident that the first varieties to occur in cultivation were the results of natural crossing with other species. All the modern large-fruited varieties examined are octoploid, and it is quite clear that they arose from crossing with the other octoploid species, F. virginiana. Great numbers of improvements, like this in strawberries, have been found to occur following the introduction of species into cultivation, and. they have always been attributed to cultivation. If "cultivation" means "opportunity for hybridisation," the popular idea is correct. No other effect of cultivation has ever been shown to have any importance in plant improvement.

It is curious that in spite of improved varieties of strawberries being raised from crossing, they were not sufficiently improved to suggest the advantage of further crossing. The only crosses, as we have seen, were accidental. In fact, we may say they could hardly have been helped. It was only a hundred years later, with the fashion set by Knight in England of raising plants from seeds, that new strawberry seedlings appeared. Soon after, by systematic crossing of the derivatives of these early seedlings, Laxton produced the first strawberries having the qualities required in cultivation to-day.

The same kind of change is undoubtedly responsible for the improvement of the Chilean wild potato. But here part of the improvement was probably carried out by the natives in South America before introduction into Europe, and only part after. In this way it is clear that almost as rapid improvement can be made by crossing polyploid species with the same chromosome numbers as by crossing diploids and raising polyploids from them. The result, too, will not depend on chance, for polyploidy is there to start with. When two tetraploid species are crossed, four different chromosome complements originally derived from four different species are thrown into the melting pot.

Not only is the result often fertile, which might well not be, were the parents diploid, but an enormously increased number of combinations can be made from the sorting out of the chromosomes derived from the four ancestral species. Among these combinations the breeder can select the new and improved forms that he wants, and propagate them vegetatively as in the case of strawberries and potatoes; or, by inbreeding, he can extract new strains that will breed true, as in the cereals.

TRIPLOIDY

So far, we have dealt chiefly with examples in which polyploidy has given advantageous results, but we must now mention some of the difficulties that polyploids offer to the breeder—difficulties which may, however, be guarded against by reference to a list of chromosome numbers.

When polyploids with the same numbers of chromosomes are crossed, the result is a hybrid which, as a rule, is of moderate fertility; but when their chromosome numbers are different, difficulties are to be expected at reduction with consequent loss of fertility. Take the case of a cross between a diploid with a single set of chromosomes in its gametes (like Prunus cerasifera, the myrobalan plum) and a hexaploid with three sets (like the "Jefferson" variety of P. domestica), such as that made by Crane.1 Irregularities of reduction might be expected. But four chromosome sets are brought together in the hybrid just as in the cross between Tritkum durum (4.x) and T. polonicum (4.x). Three of these are derived from one parent and one from the other, instead of two from each. Does this matter, since in any case the four sets are originally derived from four different species ? So far as behaviour of the chromosomes is concerned it certainly does not.

The hybrid behaves just like any other tetraploid ami even gives seedlings when back-crossed with the hexaploitl plum patent. The same kind of result has been achieved in Papaper, Crepis and Digitalis hybrids. This is well enough where there is an even number of sets of chromosomes in the hybrid, but where the number is odd, reduction can never be regular; chromosomes of the odd set have nothing corresponding to pair with, and they pass at random to one pole or the other. The result is the production of germ-cells with all kinds of chromosome numbers, and we know from other instances with odd chromosomes what this means.

Here the case is worse, for a triploid ornamental cherry like Prunus avium var. nana (a cross between the diploid sweet cherry and the tctraploid sour cherry) has eight extra chromosomes. It produces germ-cells with every number between eight and sixteen, and among these the balanced ones with eight and sixteen only come once in over a hundred times. The result is that such a triploid is highly infertile. This, however, has its uses. The triploid watercress is sterile. It is eaten in winter and possibly is sturdier on account of its seed-sterility. The cherry in question is an ornamental variety, and if the cherries of the Japanese group are examined, it is found that nearly all of them are either triploid or double-flowered. Both these types are sterile. Clearly, gardeners have chosen for ornamental purposes those varieties which produce least fruit, for in trees which are grown for a display of flowers, fruit is often a disadvantage.

Therefore, if one is raising ornamental trees which can be propagated readily by cuttings or grafts, tetraploids should be crossed with diploids to get, not merely something new, but also something with the advantage of sterility. Actually the triploid Japanese cherries have arisen accidentally, and not by design. Their sterility depends simply on the presence of an odd set which has been Introduced without hybridisation; the odd set is the result of a germ-cell with the unreduced diploid number on one side fusing with a haploid germ-cell on the other. This is what has happened in the garden tulips. Several important varieties such as Keizerskroon, Massenet, Pottebakker and Pink Beauty, are triploid. They arose from the fusion of a diploid egg with a haploid pollen grain. Consequently, unlike the products of hybridisation with non-giant tetraploid species, such triploids usually show a suggestion of gigantism which has led to their selection.

This is best exemplified by the hyacinths, but it must be pointed out that the hyacinths are altogether exceptional in not being badly affected by having odd chromosome numbers. Therefore their triploids are not sterile, but give rise to seedlings having various numbers, which, although less vigourous than the triploids themselves, are occasionally worth growing. The best varieties of garden hyacinths are triploids. The first to appear was Grand Maitre in 1870. They are all larger in bulb and flower than either the diploids which preceded them or the odd-numbered forms to which they have given rise as seedlings. They proliferate freely, their bulbils maturing in four years, that is more rapidly than those of the diploids.

SECONDARY POLYPLOIDY

The apple and pear group offer another interesting peculiarity in their chromosome properties, and therefore in their inheritance. It will be seen that the basic number 17 is not paralleled in any of the other examples that have been given. The question arises, Is this really a basic number, or does it conceal the evidence of a more complicated origin from a lower basic number ? Close study of germ-cell formation in the apples and pears and medlars has shown that 17 is not the primary number, but is really derived from an earlier set of seven, like that found in roses and strawberries. This earlier set at a remote period was doubled in part and triplicated in the other part. If the earlier set of 7 was ABCDEFG, then the new set of 17 is AAA9 MB, CCC, DDy EE, FF9 GG. It is unbalanced relative to the old set of seven, and it may be that the apple group owes its special characteristics, more particularly its fruits, to this change in balance.

Another species which probably arose as an unbalanced polyploid and owes its special qualities to its new balance is Dahlia Merckii. This species differs from other dahlias in having thirty-six chromosomes instead of thirty-two or sixty-four. It is not in the eight-series, but has two extra pairs of chromosomes. It differs in form from all the other Dahlia species with multiples of eight by having fertile ray florets. It is another case of unbalance being a good balance.

CONCLUSION

We have attempted to find, for the guidance of the plant-breeder, the materials that transmit the hereditary qualities of the plant from one generation to the next; we have concluded that these materials lie in the chromosomes. Study of the behaviour of the chromosomes has shown how their constancy is preserved and how they vary.

Particularly we see that in their reduction the chromosomes provide the mechanism of Mendelian segregation and at fertilisation the mechanism of Mendelian recombination. But in the various irregularities of their behaviour are to be seen the means by which types of inheritance can be developed much more complicated than any expressed in the simple Mendelian scheme, although they actually arise out of the same mechanism. Among these are the examples of true-breeding hybrids, such as the ring-forming evening primrose and the hexaploid bread wheat.

We also know now how sports arise from the presence of an extra chromosome and how the ever-sporting habit may be preserved by the loss of a piece of chromosome. We can see how many important improvements in plants were brought about by polyploids. We know how such plant breeders as Thomas Laxton and Thomas Andrew Knight got the results they did, although working relatively in the dark.

And we can thereforeimitate them in other fields which appear, in the light' of a knowledge of the chromosome numbers, as promising as those they worked in, The occasional seedlings of sterile diploids can be raised in order to obtain fertile polyploids. Where a sterile hybrid is no disadvantage diploids can be crossed with polyploids. We can see, moreover, that chromosome studies are particularly instructive in relation to fertility. Generational sterility is merely a symptom of segregation.

It is due to the pairing and separating of chromosomes that are not sufficiently like one another (as in the diploid hybrid Primula kewnsis), or to a less extent to chromosomes that are like one another pairing and failing to separate (as in tetraploid tomatoes). Further, there are many plants that are sterile, not on account of any hybridity in the ordinary sense, but simply because one of their parental germ-cells had two sets of chromosomes instead of one, so that they have an odd set which cannot pair normally at reduction. The degree of sterility to be expected in a plant can therefore be calculated to some extent from the pairing of its chromosomes and the pairing can be predicted from a know-, ledge of the parents' chromosomes. The plant-breeder therefore is in a position to decide before making a cross whether the results are likely to be fertile or sterile, and this is of great importance in the work of raising new econo-mic plants. Finally, we can understand a great deal more of the nature of species, a knowledge of which is as important to the plant-breeder, from one point of view, as it is to the systematise from another. A study of the chromosomes shows at once that there are some species which are homosygous, uniform and true-breeding. These are the simple-pairing diploids.

There are others, equally true-breeding and uniform, which are of hybrid origin and have certain properties, more or less clearly shown; of hybrids. These are the ring-forming diploids and hybrid polyploids. There are others again that contain several inter-sterile groups owing to polyploidy having arisen within them without hybridisation. And finally there are some that are seed-sterile hybrids maintaining themselves as uniform types by vegetative means such as bulb formation, apomixis, and so on. Cytological examination at once shows which of these groups a species falls in, and is a necessary preliminary to any systematic breeding work. While chromosome studies therefore do not give the plant-breeder any greater control over his material, they enable him to direct his efforts into the right channels for obtaining and preserving the results likely to prove most profitable to him.

NURSERY AND GREENHOUSE TECHNIQUES

All Populus propagules being commercially planted in agroforestry or in reforestation/ afforestation are E.T.P.s (Entire Transplants). These are one-year-old rooted cuttings. The tree improvement programme must therefore use E.T.P.s for clonal field trials. Nursery management techniques include soaking the cuttings and nursery beds in water before planting and then planting the 18-25cm length cuttings at a spacing of 80cm x 60cm. Debudding to remove young limbs on the E.T.P.s is done from June to November. The nursery beds are flood irrigated 1-2 times per week during March-June, until the rainy season begins. Backpack sprayers are used to apply insecticides for control of leaf beetle, thrips, and wood-borer insects.

The target E.T.P. for production nurseries is a plant whose ground-line diameter is 1/100th of the height (in cm) (personal communication with Dr. J. P. Chandra). When harvesting the one-year-old E.T.P.s in December-February, the tap root is cut at a 25cm depth and all side roots more than 10cm long are trimmed. The plants are packed, transported, sold, and planted with a naked root. Research nurseries have the additional task of developing uniformity among E.T.P.s of the same clone. Re-propagating E.T.P.s annually from the previous year's E.T.P.s for 4-5 years in the nursery may be required to remove "C-effects" (age, position-in-tree, or vigor effects on cuttings taken from the crowns of different-aged ortets on different sites).

The clones will not be taken to the field trials until within-clone variation is less than 10-15% (personal communication with Dr. Khurana at Solan). Furthermore, the multiplication of seedling-derived clones will take at least three years in the nursery to produce enough E.T.P.s for field trials at 2-3 sites. When seeds are produced from controlled crosses during breeding, they must be germinated and grown before they can be vegetatively multiplied as E.T.P.s. P. ciliata seeds are very fragile and will lose viability in 7-10 days if not germinated quickly (personal communication with Mr. D. V. Negi).

All Populus species seedlings are vulnerable to sun scald (heat) damage and diseases (damping off) during germination. Procedures for germinating and growing seedlings at WIMCO Seedlings Ltd. provide a working model for success. Fungicide-treated seeds are planted in rows on moist, heat-sterilized sand in flat clay pots and placed in a double-walled polygreenhouse with fans. The polygreenhouse is covered with 50% shade cloth. Approximately 15 days after the seeds have germinated, the germinants are 'pricked' and transplanted to individual cells in container racks. The soil in the containers is a barnyard mixture of manure that has been sterilized, so no additional fertilization is required. Containerized transplants are placed in a mist chamber in the greenhouse for one week (mist for 1-2 seconds at 5-minute intervals), then moved into a shadehouse for one week and subsequently outside under the shade of trees for 15 days. Finally, the acclimated seedlings are placed in the open sunlight where they remain until they are 30cm tall. At this size the roots can retain soil in a 'plug' when removed from the container. The 'plug' seedlings are then planted in the research nursery (personal communication with Dr. S. Chauhan of WIMCO).

AGROFORESTRY

The entire agroforestry industry with Populus is founded on a very limited genetic base. Approximately 90% of poplars being planted for agroforestry in U.P., Haryana, and Punjab come from the P. deltoides clones G-48, G-3, and S7C15 (personal communication with Dr. J. P. Chandra of WIMCO Seedlings Ltd.). The next two clones in popularity are 'Udai' from WIMCO and L-34 from the U.P. Forestry Department. Both of these clones come from open pollinations between G-48 and G-3. Other clones that are not quite as desirable, but which are being held in reserve in case of a serious disease or insect outbreak, are S7C8, S7C4, L-43, ST-240, ST-70, and ST-67. All of these clones (except the three ST clones) came from gene pools in Brazos County, Texas. The three ST clones came from the southern Mississippi River alluvial plain just north of Vicksburg, Mississippi USA.

Typical agroforestry methods are to plant 4m-5m tall E.T.P.s in (1) "bund" (shelterbelt) plantings on the small levees around flood-irrigated agricultural fields, (2) "block" plantings at 5m x 5m, 5m x 4m, or 6m x 4m spacings within agricultural fields (underplanted with the crops), or (3) "row" plantings that contain alternating rows of Poplars and horticultural tree species underplanted with agricultural crops. Typical spacing between "rows" is 6m, so the P. deltoides rows are 12m apart. Rotation lengths for poplars are 5-8 years (5 years for plywood, 8 years preferred for veneer and matches). Typical crop combinations with "bund" plantings are sugarcane or rice, but rice is not recommended because of the requirement for summer flooding.

Crop combinations with "block" plantings involve sugarcane for the first two years and then winter wheat or a combination of winter wheat alternating with summer-grown tumeric or pearl millet during the third through eighth years of the P. deltoides rotation. A new alternative to tumeric or pearl millet is celery. Dr. N. B. Singh has also observed rice being cultivated in some block plantings during the fourth through eighth years. Apparently, P. deltoides can tolerate flooding at the older ages. "Row" planting mixtures in agro-horti-forestry systems are based on a rotation length of up to 50 years for the final horticultural orchard.

These systems may involve poplar-peach, poplar-litchi, or poplar-mango combinations with sugarcane, tumeric, and winter wheat as under crops. Up to two 5-to 8-year rotations of Populus may be grown before the fruit trees (particularly mango) spread in crown diameter to become a full orchard. In all of these systems the poplar trees are pruned of lower limbs (recommended lower 1/3 of stem height after the second year). The leaves are collected and composted at the end of the growing season for placement back on the fields, and the stumps are dug up and removed for fuelwood/charcoal at the end of the rotation.

SELECTION OF THE TREE IMPROVEMENT

Tree improvement relies on understanding and using variation that naturally occurs in tree populations. Tree improvement increases the value of a tree species by 1) *selecting* the most desirable trees from natural stands or plantations, 2) *breeding* or mating these select trees and 3) *testing* the resulting progeny. The trees involved is this process are referred to as the*breeding population.* This three-step process is then continuously repeated to further improve the average value of the breeding population. Each iteration of this improvement process is referred to as a *generation*.

The *production population* is another group of trees established to meet commercial planting demands. Usually the production population is a *seed orchard* or group of trees at a single location managed specifically for seed production. For species which can be readily propagated using rooted cuttings, a *hedge orchard* managed to produce cutting material may serve as the production population. Seed and hedge orchards generated from a tree improvement programme are established from grafts, rooted cuttings or seed of the very best trees in the breeding population.

SELECTION

Tremendous variation exists in natural stands for many important Christmas tree characteristics such as growth rate, colour, branching habit and disease resistance. Growers often encounter this variation in their Christmas tree plantations. Observed characteristics such as growth rate are referred to as the *phenotype* of a tree. A tree's phenotype is determined by both its genetic constitution, or *genotype* and the effect of the *environment* as described by the following equation:

$$P = G + E$$

where,

P = phenotype

G = genotype

E = environment.

When selecting trees, what you see is not necessarily what you get. Genetically inferior trees may sometimes appear phenotypically desirable because they grew in an unusually favorable micro-environment. Conversely, genetically superior trees may appear phenotypically undesirable due to poor environmental conditions. Characteristics vary in the degree of genetic versus environmental influence. For example, genetic control of branching habit is stronger than for height growth so that the selection process is generally more effective for branching traits. Initial selection of trees from natural stands or unimproved plantations are necessarily based solely on phenotype. Since the variation due to environmental effects is less in plantations than natural stands, selection is more effective in plantations.

In order to ensure that the selection process is as effective as possible in unimproved stands, candidate trees are graded according to predetermined criteria. Both the candidate tree and neighboring check trees are measured and compared in order to account for local environmental effects. Once a tree improvement programme is underway, genetic information (performance of relatives) is used to increase the effectiveness of the selection process.

BREEDING

The next step in the tree improvement process is mating or breeding among the select trees. For this purpose, branch tips or *scion* of each selection are grafted onto seedling *rootstock* to establish a *breeding orchard. Control-pollinations* are then performed among the selections. Control-pollination starts with covering receptive female cones with bags during early spring to prevent natural pollination. Pollen collected from other select trees or mixed from a group of select trees is then injected into the pollination bag. After pollination has occurred, the bags are removed and the cones allowed to ripen. Each cone-producing branch is labeled to ensure correct identification at cone harvest. Control-pollination is performed according to certain*mating designs* among the select trees in order to maximize the overall genetic information derived from the resulting progeny.

TESTING

Seed produced from tree improvement breeding efforts are used to establish progeny tests. The purpose of these tests are to 1) provide genetic information about the select parent trees and 2) provide an improved population of trees from which the next generation of select trees is made. Tree families established in progeny tests are randomized and replicated in a designed manner to meet statistical and genetic criteria. Families are planted in several sites and in more than one year to sample an adequate number of environmental conditions. Christmas tree progeny tests are intensively managed similarly to a typical Christmas tree plantation. Growth and quality measurements are made periodically in these tests until harvest. These data are entered into a data base and analyzed. The results are used to assess the genetic worth of the original selections, to make selections for the next generation and to make recommendations for establishing and upgrading seed and hedge orchards.

Final Remarks

Tree improvement is the successive application of selection, breeding and testing to continuously improve the value of a population of trees. The best trees of the population are used to propagate planting stock for commercial plantations. While the Christmas tree industry stands to reap tremendous benefits from tree improvement, it should not be viewed as a panacea. Tree improvement is one of many tools needed to continue to improve the productivity and quality of Christmas trees crops. Growers must continue to use the best available technology to manage their plantations. In fact, the greatest benefits from tree improvement will be achieved on the best sites receiving the best management.

Due to the nature of trees, the tree improvement process requires both a large-scale effort and a relatively long time-frame to achieve its results. Researchers, extension personnel, government and commercial propagators, and growers must all cooperate to ensure that these results are realized as rapidly as possible. While hard work and patience will be required, the Christmas tree industry will be rewarded with new standards of productivity, quality and profitability.

6

Periodicity and Pattern in Individual Tree Growth

Tree growth is periodic over short time-spans and follows a definite pattern in the long term. In temperate countries growth is overwhelmingly determined by the season, with growth confined to a period of a few weeks to perhaps several months in any one year while warmth and moisture are adequate. Even in the moist tropics, with year-round favourable growing conditions, trees show periodicity under a measure of genetic control as recorded, for example, from measurements of multinodal tropical pines such as *Pinus caribaea* and fast-growing hardwoods such as *Cordia alliodora.* As well as periodicity in one year, trees exhibit a strong pattern of growth during their life. In relation to age the pattern is typically sigmoid, whether recording total height, diameter or volume over time.

DYNAMICS OF STAND GROWTH

The dynamic nature of stand growth is a little less obvious than for a single tree, since superficially a stand seems to be a collection of individual trees the relations of which appear constant. That this is not so is seen by considering a small tree in a mature even-aged stand. At regeneration (or planting) this small tree will have been about the same size as its neighbours but subsequently will have competed less successfully for light, nutrients or moisture to become inferior to its neighbours.

As a stand develops any apparent uniformity disappears, with some trees developing vigorously and neighbours becoming suppressed, moribund and even dying. Thus the stand is continually changing - it is dynamic. It is not only a collection of individual trees, each with its own genetic potential for using the site, but also a collection of trees that interact and compete with one another.

Seedlings and Regeneration

From the outset trees appear to grow at different rates. Even in new plantations of well-spaced trees of one species, and long before onset of between-tree competition, the growth of individual trees is not identical. Moreover, there is often no correlation between initial postgermina-tion vigour and subsequent growth rate. Small seedlings, or plants from the forest nursery, do not necessarily lead to small slow-growing trees.

Between-tree Competition

As trees grow they eventually begin to compete with their neighbours. This has both an above-ground component, mainly competition for light, and a below-ground element in terms of root competition for nutrients and moisture. Such competition is most readily seen in suppression of side branches on the lower crown and a measurable impact on diameter increment.

The timing of the onset of between-tree competition depends on distance between trees, but is usually first measurable in forest stands from about the time of canopy closure. The principal exception is in arid climates where competition between root systems for moisture greatly exceeds that for light, and trees and shrubs remain widely dispersed and rarely in contact above ground.

Differentiation into Crown Classes

As competition begins, growth of some trees slows more than others. These tend to be the smaller trees at the time of canopy closure and the competition reinforces their inferior status. Once dominated, few trees can recover unless a gap develops owing to death of a neighbour. Thus as a stand develops, a range of tree sizes emerges and, traditionally, these are classified into different crown classes according to the tree's relative position in the canopy.

Interventions and Manipulations

These empirical relationships of how trees are observed to grow and how they develop and interact in a stand provide the basis for manipulating their behaviour. Adjusting spacing between trees, both when planted or through thinnings, or influencing the balance of a mix of species in a stand all impact on how a stand develops, on how the increment is distributed and hence on the composition of tree types, trees sizes and the total of woody growth that will result.

In summary, densely stocked forest leads to high volumes per unit area but small mean tree size compared with less well-stocked forest on a similar site and of the same age. In the latter case there will be fewer but larger diameter trees, although total volume of timber may be somewhat reduced. This allows forests to be managed in different ways to yield different assortments of products. How to model these relationships and outcomes in detail, beyond the purely empirical, forms the bulk of this chapter.

GROWTH MODELS

The history of forest growth models is not simply characterized by the development of continuously improved models replacing former inferior ones. Instead, different model types with diverse objectives and concepts were developed simultaneously. The objectives and structure of a model reflect the state of the respective research area at its time and document the contemporary approach to forest growth prediction. The history of growth modelling thus also documents the advancement of knowledge in the science of forest growth.

Beginning with yield tables for large regions as a basis for taxation and planning, model development led to regional yield tables and site-specific yield tables and culminated in the construction of growth simulators for the evaluation of stand development under different management schemes. Vanclay (1994) provided an overview about growth and yield management models and their application to mixed tropical forests. The 1980s brought a new trend with the development of ecophysiological models, which give insight into the complex causal relationships in forest growth and predict growth processes under various ecological conditions. The emphasis in model research has shifted towards ecophysiological models and away from models aimed only at providing growth and yield information for forest management. These models attempt to simulate forest growth on the basis of fundamental ecophysiological processes. The scientific value of ecophysiological models cannot be overrated; however, they will not be applied in forest management for the next few years as they are in many ways not yet sufficiently validated. Also, input and output variables do not yet meet the demand of forest management practice.

A major change has taken place in model conception, *i.e.* the understanding of forest growth on which the model is based. The tables by Weise (1880), Schwappach and Wiedemann resulted from a purely descriptive analysis of sample area data in the form of total and mean values of observed processes of stand development. These descriptions were later combined with theoretical model concepts that also considered natural growth relationships and causal relations as far as they were known at the time.

For example, yield tables for mixed stands of pine and beech created by Bonnemann (1939) characterize growth of beech in the middle and lower storey by mean values. The FOREST model of Ek and Monserud (1974) controls increment behaviour of lower-storey trees by geometrical competition indices, and the ecophysiological growth models of Bossel (1994), Mäkelä and Hari (1986) and Mohren (1987) derive increment behaviour of lower-storey trees from light availability and performance in terms of photosynthesis. The change in model objectives and concepts is closely related to a change in quality of the information generated. Pure management models aim at reliable prediction of forest yield values that are crucial for planning and control in forest management, *e.g.* height and diameter increment and associated economic value.

Ecophysiological models aim at biomass development, nutrient input and loss, etc.; variables relevant to forest management are only of secondary importance in these models. For future planning in modern forestry, models meeting the information demands of ecology as well as of economy will gain in importance. Ecophysiological models and stand management models can give specific decision support. Ecclogical and socioeconomic conditions define the framework and thus the 'foundations' for management decisions. Ecophysiological models can support the ecological elements of the framework, for example the effect of site conditions, species mixture and thinning variants on critical loads, water quality or acidification. Stand treatments of interest can be judged in this way as ecologically acceptable or unacceptable. Management models help to optimize the path from starting point to objective via the given framework, for example they support the decision between different thinning and pruning strategies.

With the shift from tree and stand management models with low resolution to more complex ecophysiological models, different source data are needed for model construction and for the determination of model parameters. Standard datasets derived from research sample plots (diameter, height, etc.) were used for the development of stand growth models for applied forestry. For the construction of single-tree models, additional data are required (crown dimension, tree position, etc.). The transition to ecophysiological models requires an additional database that can only be provided by broadening experimental concepts and cooperation with neighbouring disciplines.

Models are always an abstraction of reality and are greatly influenced by the modeller's knowledge and perception of nature. This applies to the construction of yield tables as well as for eco-physiological models.

Stand Growth Models based on Mean Stand Variables

With a history of over 250 years, yield tables for pure stands may be considered the oldest growth models in forestry science and forest management. They are representations of stand growth within defined rotation periods and are based on a series of measurements of diameter, height, biomass, etc. reaching far back into the past. From the late eighteenth to the middle of the nineteenth century, German scientists such as Paulsen (1795), von Cotta (1821), R. Hartig (1868), Th. Hartig (1847), G.L. Hartig (1795), Heyer (1852), Hundeshagen (1825) and Judeich (1871) created the first generation of yield tables based on a restricted dataset. These original yield tables soon revealed great gaps in scientific knowledge. A series of long-term data-collection campaigns on experimental areas was therefore started. This was the birth of a unique network of long-term experimental plots in Europe that is still under survey.

The second generation of yield tables, initiated towards the end of the nineteenth century and continued into the 1950s, follows uniform construction principles proposed by the Association of Forestry Research Stations (the predecessor organization of the International Union of Forest Research Organizations (IUFRO)), in 1874 and 1888 and has a solid empirical basis. The list of protagonists involved in this work includes Weise (1880), von Guttenberg (1915), Zimmerle (1952), Vanselow (1951), Krenn (1946), Grundner (1913) and, in particular, Schwappach (1893), Wiede-mann (1932) and Schober (1967), who designed yield tables that were conceptually related and is still being used to this day. A brilliant example of their work is the yield tables for European beech. In the 1930s and 1940s the first models of mixed stands were contructed under the direction of Wiedemann. Data from some 200 experimental areas established by the Prussian Research Station led to the yield tables for even-aged mixed stands of pine and beech, spruce and beech, pine and spruce, and oak and beech. The Second World War prevented Wiedemann from bringing the development of yield tables for uneven-aged pure and mixed stands to an end, but his studies initiated systematic research on mixed stands. Yield tables for mixed stands of this generation were never consistently used in forestry practice as they were restricted to specific site conditions, intermingling patterns and age structures.

Yield tables developed by Gehrhardt (1909, 1923) in the 1920s effected a transition from purely empirical models to models based on theoretical principles and biometric formulae and led to a third generation of yield tables. These models were designed by, among others, Assmann and Franz (1963), Hamilton and Christie (1973, 1974), Vuokila (1966), Schmidt (1971) and Lembcke *et al.* (1975), and at their core is a flexible system of functional equations. These functional equations are based as far as possible on natural growth relationships and are generally parameterized by means of statistical methods. The biometric models are usually transferred into computer programmes and predict expected stand development for different spectra of yield and site classes. A wealth of data were available for the construction of these models and processed with modern statistical methods.

Since the 1960s a fourth generation of yield table models has been created, *i.e.* the stand growth simulations of Franz (1968), Hoyer (1975), Hradetzky (1972), Bruce *et al.* (1977) and Curtis *et al.* (1981, 1982), which simulate expected stand development under given growth conditions for different stem numbers at stand establishment and for different tending regimes. They describe stand development at different sites and for varying treatments and varying numbers of trees at the time of establishment. Expected stand development under given growth conditions is simulated by means of computer programmes and controlled by systems of suitable functions forming the core of the growth simulator. All information available on forest growth is synthesized into a complex biometric model that simulates stand development for a wide range of possible management alternatives and summarizes the results in tabular form similar to yield tables. These yield tables reflect the stand dynamic for a wide range of imaginable management scenarios. While table and model were identical for the yield tables of earlier generations, simulator-created yield tables now describe just one of many potentially computable stand development courses.

Despite a number of drawbacks, yield tables still form the backbone of sustainable forest management planning. When computing capacities and available data for model construction increased and with the rising demand for information in forestry, mean-value and sum-orientated growth models and yield tables were increasingly replaced by stand-orientated growth models, predicting stem number frequencies, and by single-tree growth models. Prodan (1965, p. 605) commented on the significance of yield tables in the context of silviculture and forest sciences: 'Undoubtedly, yield tables are still the most colossal positive advance achieved in forest science research. The realization that yield tables may no longer be used in the future except for more or less comparative purposes in no way detracts from this achievement'.

Stand-orientated Management Models Predicting Stem Number Frequency

With the transition towards new intensive treatment concepts, the demand for information in forestry has changed the emphasis from mean stand values towards single-tree dimensions of selected parts of a stand. This changed demand for information resulted in the 1960s in the creation of the first growth models, which enabled prediction of mean stand values as well as frequencies of single-tree dimensions. Until then, a stand served as the usual information unit on which all predictions were based; these predictions were now strengthened by statements about stem number frequencies in diameter classes, which are needed for precise prediction of assortment yield and value of a stand. Depending on their concept and construction, stand-orientated growth models predicting stem number frequency are classified into differential equation models, distribution prediction models and stochastic evolution models.

Many natural processes in various disciplines of the natural sciences can be described by differential equations. Examples are the differential equations formulating change of yield descriptors for diameter classes of a stand, *i.e.* change of stem number, basal area and growing stock, depending on current yield state values. Stand development then results from the numerical solution of the differential equations. In the 1960s and 1970s, Buckman (1962), Clutter (1963), Leary (1970), Moser (1972, 1974) and Pienaar and Turn-bull (1973) developed stand-orientated growth models based on differential equations.

In the mid 1960s, Clutter and Bennett (1965) proposed a completely new approach to stand growth modelling. They characterized the condition of a tree population by its diameter and height distribution and described stand development by extrapolation of these frequency distributions. The precision of such models is decisively determined by the flexibility of the distribution type on which it is based. The suitability of different distribution types, for example beta, gamma, lognormal, Weibull or Johnson, has to be assessed individually. Compared with those reviewed earlier, in these models stand development is not controlled by the age function of the individual yield descriptors but by the parameters of the underlying frequency distribution. Models of this type were initially constructed by Clutter and Bennett for North American spruce stands and further developed by McGee and Della-Bianca (1967), Burkhart and Strub (1974), Bailey (1973) and Feduccia *et al.* (1979).

The term 'evolution models' for stochastic growth models is derived from the fact that in these models stand development evolves from an initial frequency distribution, for example from a diameter distribution known from forest inventory. Thus these models, like distribution prediction models, predict frequencies of single-stem dimensions. However, the mechanism accounting for the extrapolation is based on a Markov process, giving the transition probability for the shift between the diameter classes. Stochastic growth models were introduced to forestry science with the pioneering investigations by Suzuki, and they continue to be linked to his name today. His growth models, for Japanese *Chamaecyparis* pure stands for example, have been consistently developed by Sloboda (1976) and his team since the mid 1970s; they are mainly interested in adapting the models, which are orientated to Japanese conditions, to the issues of German forestry and in model validation based on permanent test plot data. Stand-orientated growth models based on stochastic processes have been developed by Bruner and Moser (1973) and Stephens and Waggoner (1970) also for mixed stands.

Single-tree Orientated Management Models

Single-tree models describe the stand as a mosaic of single trees and model individual growth and interactions with or without consideration of tree position. This has paved the way for the design of models of pure and mixed stands of all age structures and intermingling patterns. An equation system that controls growth behaviour of single trees depending on their constellation within the stand is the central module of all single-tree models. Position-independent or position-dependent competition indices are used to quantify the spatial growth constellation of each tree and to predict its increment of height, diameter, etc. in the following period. Compared with stand-orientated growth models based on mean stand descriptors and those predicting stem number frequencies, single-tree models work on higher resolution. The information unit in singletree models is the individual tree. However, results of lower-resolution models, for example mean tree development or diameter frequency distributions, can also be derived from single-tree model results by integration. Information about stand growth then results from summarizing and aggregating each individual single-tree development for a given growth period. Recent single-tree models are programmed to enable the user to influence a simulation run interactively. This allows stand development to be followed step by step during the simulation and permits the user to specify other factors (*e.g.* thinning or influence of disturbance) at any time during the simulation process, thus influencing or diverting the current course of stand development.

After parameters for the control of the singletree model have been set, tree characteristics at the beginning of the prediction phase for the test area to be investigated are fed into the computer as initial values for the simulation. This tree list can contain data on tree species, stem dimensions, crown morphology, stem position and other data about the stand individuals. These data usually originate from single-tree-based inventories of indicator plots. Starting with these initial values, change (*e.g.* mortality or development of diameter, height or crowns) for all stand members depending on individual growth conditions is predicted using an appropriate control function; this is done for a first growth period, for example 5 years. Once the tree list has been processed, change of growth conditions (*e.g.* due to thinning or disturbance) can be specified prior to continuing to the next increment period.

This will now influence single-tree growth in the following period. The modified state values of all trees resulting at the end of the first growth period also represent the initial values for the second growth period. These values are repeatedly extrapolated in every simulation cycle and interim results are given. The simulation continues until the envisaged prediction period has been completed step by step. In most models, time steps are 5 years, sometimes only 1 or 2 years. By removing single trees during a simulation run, the growth constellation and growth behaviour of the remaining individuals change in the next growth period. Growth reaction of the stand is thus explained by the reactions of all single trees to this intervention. By relating stand development back to growth behaviour of single trees and by modelling single-tree dynamics depending on growth constellation within the stand, single-tree models, after being initialized accordingly, enable evaluation of a wide range of treatment programmes.

The first single-tree model was developed for pure Douglas fir stands by Newnham (1964). It was followed by the development of models for pure stands by Arney (1972), Bella (1970) and Mitchell (1969, 1975) and colleagues. In the mid 1970s, Ek and Monserud applied the construction principles for single-tree orientated growth models for pure stands to uneven-aged pure and mixed stands (Ek & Monserud 1974; Monserud 1975). Munro (1974) distinguished distance-dependent and distance-independent single-tree models, the former being able to refer to data about stem position and stem distance for the control of single-tree growth. The worldwide bibliography of single-tree growth models compiled by Ek and Dudek (1980) lists more than 40 different single-tree models, which are grouped into 20 distance-dependent and 20 distance-independent models. Single-tree models developed since the 1980s in many ways go back to the methodological bases of their predecessors; however, owing to the rapidly improving technology of modern computers they are far more user-friendly than older single-tree models.

ECOPHYSIOLOGICAL GROWTH MODELS

All the models mentioned above rely on growth and yield data from long-term observation plots and hence have the advantage of being validated empirically. However, there is a drawback to historically deduced data in as much as growth conditions undergo changes, and reaction patterns from the past cannot simply be projected into the future. In the 1970s, model research was pointed in a new direction with the creation of high-resolution ecophysiological process models, which account for metabolism, organ formation, assimilation and respiration as well as biochemical and soil chemistry reactions. Pioneers of the ecophysiological process model for forest stands are Bossel (1994), Mäkelä & Hari (1986) and Mohren (1987). The term 'process model' is slightly misleading in the sense that all forest growth models describe processes. Only the temporal and spatial scales of modelled processes become more detailed and accurate in the transition from yield table models via single-tree management models and succession models to growth models based on ecophysiological data.

The development of modern process models begins with a systems analysis and the selection of characteristic system components. A system to be analysed and modelled is first described using methods of systems analysis. Results of this description can be transferred into a system diagram. The description breaks the system down into system components characteristic for all biological systems and identified by different symbols in the system diagram.

By system parameters we mean those that remain constant during the lifetime of the system. Exogenous parameters are variables that control the system but which cannot be influenced by the system, *e.g.* stress caused by air pollutants. State variables are the actual output value of the model; their current values reflect the system's state. Important state variables in stand models are accumulated carbon quantities in needles, branches, stem and roots. The initial values of the state variables give the starting values of a system and thus crucially influence its further development.

In a growth model for example, stem number and initial stand structure have to be specified as initial values. The rate of change of the state variables controls change, *i.e.* input and output of state variables. Examples are mortality rates or respiration rates, which control the change of the carbon quantities accumulated in the different components. Intermediary variables change simultaneously with the state variables and feed back into the system. The system components are indicated in the system diagram with different symbols and their interrelations are identified by arrows. The model thus outlined is transferred into a mathematical model and subsequently into a computer programme. For this, the system components and links are described by mathematical or logical relationships. Once the complete model is constructed, the causal relations implemented are parameterized. The system behaviour can be simulated with the developed computer programmes. All suitable information known about the system is therefore consolidated in the system components and the system structure. The process of system analysis and model development concludes in the validation of the final model. For validation, *i.e.* testing if the causal relations assumed in the model realistically reflect growth of stands or single trees, empirical yield data can be used. If necessary, individual model assumptions are corrected or model parts revised.

A vastly improved understanding of ecophysiological processes in forest ecosystems paved the way for this model approach and it was the actual modelling of these processes that provided an idea of the functioning of the overall system. A further impetus to process model development was the need to understand and predict the reactions of forest ecosystems to an increasing number of adverse effects, such as industrial emissions, rise in atmospheric CO_2 and climate change. In the context of environmental instability, high-resolution and accurately detailed process models are certainly the ideal approach for understanding and predicting forest ecosystem behaviour. However, there are particular constraints in developing and applying process models due to considerable gaps in our knowledge of part processes in assimilation organs and in the soil.

Also, the scaling-up of part processes to the behaviour of the overall system is still largely unresolved. Moreover, the introduction of process models still requires intensive research and extremely high-powered computers that are only rarely available in practice. To date, process models are therefore primarily research instruments rather than forest management planning tools.

Gap Models and Biome Shift Models

In the view of modern theoretical ecology, a spatially extensive system is composed of mosaiclike subunits and can be studied by analysing these subunits. Watt (1925, 1947), Bormann and Likens (1979) and others transferred this view of extensive ecosystems to the study and model representation of the growth dynamics of pure and mixed stands. This laid the foundations for the concept of gap models suitable to predict succession. According to this concept, a forest stand is an aggregation of gaps. The size of these gaps corresponds to the extent of a potential crown area of a dominant tree or tree group (areas of 0.04-0.08 ha). The actual information unit is the tree group in the gap; stand development results as the sum of the total spectrum of contributing gaps. Gap models imply that forest development in a gap occurs in a fixed cycle: A gap results from exploitation or death of a dominant tree, and thus the growth conditions of understorey trees improve and natural regeneration occurs. Growing trees successively close the gap and a new overstorey develops. The cycle is repeated with further losses of dominant trees. Growth models using this approach were predominantly employed for investigations of competition and succession in semi-natural stands.

Gap models, such as those designed by Shugart (1984), Pastor and Post (1985), Aber and Melillo (1982) and Leemans and Prentice (1989), are primarily aimed at mixed stands. While in the models described above increment-determining factors have effects on stands or individuals respectively, gap models describe tree growth that depends on growth conditions in the individual gap. Gap models simulate growth dynamics for single trees or tree classes in a gap; it is therefore possible to generate information about the development of diameter, height and volume of single trees as well as stands.

However, regarding input and output variables they are less dependent on information available from, or required by, forestry practice; rather, they aim at predicting long-term succession in natural forest stands and the effects of altered growth conditions. The FORMIX2 model for virgin and logged Malaysian lowland dipterocarp forests is an example of an ecophysiological-based gap model with output variables that is useful as decision support in forest management.

Biome shift models, such as those of Box and Meentemeyer (1991) and Prentice *et al.* (1992), establish statistical relationships between regional climate and vegetation type. Based on relevant climatic conditions, the nature of potential biomes, *i.e.* communities, may be predicted on a regional and even global scale. Of all the models under discussion, these are the ones that provide the highest aggregation of data on vegetation development and forest growth. They have therefore gained increasing importance in research on global change.

Hybrid Models for Forest Management

The transfer of specific components of ecophysio-logical models (based on solid process knowledge) into stand or single-tree management models (based on long-term experimental plots and increment series) leads to what Kimmins called 'hybrid growth models'. Models of this type were constructed by, among others, Botkin *et al.* (1972) and Kimmins (1993). Their objective is to make the best possible use of the newly acquired knowledge of ecophysiological processes combined with historical increment observations to assist in forest planning and management. On account of the implemented relationship between site conditions and species-specific growth, they can be used for pure and mixed stands. In the past 100 years mixed stands have gradually become the focus of forest research, particularly on account of studies by Gayer (1886), Wiedemann (1939b) and Assmann (1961), but to this day growth models for mixed stands are scarcely used as quantitative planning tools. Only very recently have models created by Kolström (1993), Nagel (1996), Pretzsch (1992), Pukkala (1987) and Sterba *et al.* (1995) found use in forestry practice for planning work in pure and mixed stands. These are in effect site-sensitive single-tree models constructed from a broad base of ecophysiological and growth and yield data. Version 2.2 of the SILVA model, developed in Germany for pure and mixed stands, belongs to the category of hybrid models and may be used as an example to explain the functional principles underlying this approach.

Management Model SILVA 2.2 for Pure and Mixed Stands

SILVA reflects the spatial and dynamic character of mixed-stand systems in as much as it models spatial stand structures at 5-year intervals. This permits the recording of the individual growth constellation of every tree and the control of tree increment in relation to growth constellation and the original dimensions of the tree. The external variables determining tree increment and stand structure are treatment, risk and site factors. The model simulates the effects that tending, thinning, regeneration and natural hazards such as storms and wind have on the stand dynamic.

The feedback loop, stand structure —> tree growth —> state of tree —> stand structure, forms the backbone of the model. The step-by-step modelling of the growth of all individual trees via differential equation systems provides information about the development of assortment yield, financial yield, stand structure, stability and diversity of the stand over and above the data, required in yield calculations, on height, diameter at breast height, number of stems, etc. Input and output data used in the model correspond to the data available from, or required in, forestry practice, for example only site variables available on a large scale are considered. With models of this type a weighting between yield-related, socioeconomic and ecological aspects of stand development in pure and mixed stands becomes possible. Parametrization relies on yield and site characteristics of pure and mixed stands that have been under observation for over 100 years.

The position-dependent individual tree model SILVA 2.2 breaks down forest stands into a mosaic of individual trees and reproduces their interactions as a space-time system. It can therefore be used for pure and mixed stands of all age combinations. Primarily it is designed to assist in the decision-making processes in forest management. Based on scenario calculations SILVA 2.2 is able to predict the effects of site conditions, silvicultural treatment and stand structure on stand development, and therefore also serves as a research instrument.

A first model element reflects the relationship between site conditions and growth potential and aims at adapting the increment functions in the model to actual observed site conditions. With the aid of nine site factors reflecting nutritional, water and temperature conditions, the parameters of the growth functions are determined in a two-stage process. The stand structure generator STRUGEN facilitates the large-scale use for position-dependent individual tree growth models. The generator converts verbal characterizations as commonly used in forestry practice (*e.g.* mixture in small clusters, single tree mixture, row mixture) into a concrete initial stand structure with which the growth model can subsequently commence its forecasting run.

The three-dimensional structure module uses tree attributes such as stem position, tree height, diameter, crown length, crown diameter and species-related crown form to construct a spatial model of the stand in question. The thinning model is also based on individual trees and can model a wide spectrum of treatment programmes. The core of the thinning model is a fuzzy logic controller. In the simulation studies described below the thinning model simulates various thinning methods (thinning from below and selective thinning) and thinning intensities (slight, moderate and heavy). The competition model employs the light-cone method and calculates a competition index for every tree on the basis of the three-dimensional stand model. The allocation model controls the development of individual stand elements. Tree diameter at height 1.3 m, tree height, crown diameter, height of crown base, crown shape and survival status are controlled, at 5-year intervals, in relation to site conditions and interspecific and intraspecific competition. Finally classical yield information on stand and single-tree level for the prognosis period are compiled in listings and graphs. Additional information on stem quality, assortment and financial yield complete the growth and yield characteristic. At every stage of the simulation run, a programme routine for structural analysis calculates a vector of structural indices that serve as indicators for habitat and species diversity and form a link to the ecological assessment of forest stands.

The algorithmic sequence for predicting forest development comprises the following steps. Step 1 is the input of data on the initial structure and site conditions of the monitored stand. In step 2, the parameters of the growth functions are adapted to actual site conditions.

Once the starting values for the prognostic run are complete, monitoring can begin. If there are no initial values, for example stem positions are unknown, the missing data can be realistically complemented with the help of the stand structure generator (step 3).

Once the spatial model has been constructed (step 4) the silvicul-tural treatment programme is specified in step 5. The competition index calculated for each tree through the three-dimensional model in step 6 is used, in step 7, to control individual tree development. Steps 4-7 are repeated until the entire prognostication period has been run through in 5-year steps. To date, model research has had little success in substituting the yield tables for pure stands by an improved information system for pure and mixed stands. This can in no way be attributed to a deficit in methodological principles, data or technical equipment. Rather, the causes lie in the fact that new models are not properly adapted to practical requirements. The recent introduction of the growth model SILVA 2.2 for forest management use led to a range of operational requirements and outputs demanded from the management models that will be used in decision-making processes at stand and forest enterprise levels.

1. The natural management of forests is currently making great headway. In the long run only those growth models capable of simulating the growth of pure and mixed stands of all age compositions and structural patterns will find approval.
2. Models need to be operable at stand and forest enterprise levels and able to simulate growth behaviour under different thinning regimes and different processes of artificial and natural regeneration.
3. Flexibility of the model is essential so as to permit simulation of growth reactions to site alterations and interference factors on a large regional scale.
4. Apart from tree and stand characteristics such as volume production, assortment yield, wood quality and financial yield should also include structural parameters determining the recreational and protective functions of forests as well as indicators showing the impact of hazards or ecological instability.
5. Forestry practice is interested, first and foremost, in calculating scenarios at stand and forest enterprise levels. This can only be achieved if input and output data of the model consider what information is available and which data are needed in forestry practice. Furthermore, achieving this goal also depends on whether the model forms part of a comprehensive forestry information system and, lastly, whether hardware specifications are acceptable in practice.

For decades forestry practice has been hoping for improved growth models to assist with research, planning, operations and control in forest management. The general acceptance of new models by practitioners calls for close cooperation between forest science and forest practice, from the design and development of the model to its actual introduction in forest management.

Likewise the only capital equipment required is a weaving needle, which is a very cheap item to buy. Similarly, wood is often used to make a variety of household implements and agricultural tools, or is carved to make products aimed at the tourist market. Once again, interviews with resource makers stress that the skills used to make these objects can be acquired in a very short time and the capital equipment required is minimal. Finally, where NTFP foods are processed to make commercial products, such as fruit wines, palm wines, dried meats, dried fish and dried insects, again the skills and capital equipment required are low: it is essentially a matter of allocating sufficient household labour to complete the task. Due to the difficulties that rural households face in capital accumulation, because formal credit markets generally fail in rural areas and because education levels are low, rural households are usually endowed with unskilled labour but poorly endowed with capital. So another reason for the extensive use of NTFPs by rural households is that the factor inputs required to collect and process NTFPs (unskilled labour, little capital equipment) match closely the factor endowments of the classic rural household. (These required factor inputs, combined with open access to the NTFP resource, means that NTFP activities have very low entry barriers. In such circumstances we would therefore expect the returns to NTFP activities to be correspondingly low.)

Fourth, households also use NTFPs in response to the general riskiness of rural economic activities. The use of NTFPs, particularly what have been classified as 'minor forest products', during times of household stress is an observation common to much of the NTFP case study literature. NTFPs display a certain degree of non-covariance with respect to agricultural output. When crops fail due to drought or disease, or when shocks hit the household such as unemployment, death or disease, some NTFPs will still be available for the household to either consume or use to generate cash income to purchase its essential needs. Thus it is economically rational for risk-averse rural households to hold a portfolio of production opportunities that exploits this non-covariance. In this sense, commonly held forests and woodlands can be regarded as providing a set of back-stop resources that insure households against the failure of other, higher return production activities. Thus, even if the returns per hectare of woodland were systematically below that of agricultural land, rural villages would still retain common lands for this purpose.

Finally, the argument comes full circle. One of the reasons that rural households use NTFPs so extensively is precisely because they are open-access resources. There are a number of reasons for the survival of communally held resources, for example the insurance element; however, a major reason for the survival of such resources is the very low income levels of rural households. Resource privatization is an expensive business: households taking charge of privatized resources usually need to spend time and money on creating exclusion (*e.g.* by building fences), on enforcement (*e.g.* employing guards to monitor incursions) and on punishing infractions (by going through the courts). Particularly where a resource has previously been held in common, privatization is often strongly contested and consequently these various costs can be expected to be high. Where households are generally poor, the private costs of resource privatization are likely to be far higher than the potential private gains from control of the resource. Thus the low incomes of rural areas underpin the existence of commonly held resources, which in turn underpins the extensive use of NTFPs by those very same low-income households (on the relationship between efficient property rights regimes and transactions costs.

CAN RURAL DEVELOPMENT BE BASED AROUND NTFPS?

We have presented evidence that rural households' use of NTFPs is widespread and significant in terms of their overall livelihoods. We have also suggested that the cause of this extensive NTFP use by poor rural households is the match between the economic characteristics of NTFPs and the economic endowments and constraints of those same rural households. Can we therefore conclude from this that the commercialization of NTFPs is economically feasible and likely to bring substantial economic benefits to rural poor? Further, if commercialization were feasible, what would it be likely to do to the resource base that produces these NTFPs? In other words, what light does the analysis above shed on the possibility of using NTFPs to meet the twin goals of economic development and environmental conservation? I argue that there are substantial difficulties here, arising from the same economic characteristics of rural households and the causes of their NTFP use that were listed earlier. I look at a number of issues in turn and draw on the case study literature to illustrate where they have led to problems for various NTFPs.

The Problem of Preferences

We saw earlier that one of the reasons for rural households' NTFP use was their low incomes and the still-partial monetization of many of their economic transactions. However, it is a truism that as individuals and households become more affluent, their demands for goods and services change. In particular, people choose to consume more highly preferred goods or goods of a higher quality. (There is abundant evidence for these demand shifts in the higher-income developed countries.) However, evidence from academic studies and from history suggests that NTFPs do not have high product quality or are not in general highly preferred goods. One way of showing this is to look at income elasticities of demand for NTFPs, as these elasticities reveal what the demand response of households will be to any increase in their incomes.

Unfortunately, demand elasticities for NTFPs are not easy to come by, due to the lack of attention paid to such goods by formal economic studies. Thus we have culled evidence from two sources. The first is the author's own study of households and environmental resources in Shindi, Zimbabwe and the second is from a recent literature review reported in Köhlin (1998). Table contains income elasticities for a range of NTFPs: wild fruits, wild vegetables, wild animals, firewood, NTFP-derived goods and thatching grass. In general, the income elasticities for these goods are low. In only one case, wild fish demands, is the recorded elasticity greater than 1 (*i.e.* a luxury good), although it is believed this has much to do with the supply conditions of wild fish in the study site. Otherwise most elasticities are closer to zero than to 1. For example, wild fruits collectively and mice have income elasticities of 0.3-0.4; woodland-derived wild vegetables have an elasticity of 0-0.2; firewood has an elasticity of 0.4; and NTFP-derived goods have income elasticities of 0.4-0.5. While these elasticities are positive, so that increases in income within the range of incomes spanned by the sample will increase overall NTFP demands, the budget shares of these goods will decline, to be replaced in importance by other more preferred goods. Table gives one example of this, by comparing the income elasticities of woodland-derived and field-derived wild vegetables. The fact that the latter are much higher than the former reflects the higher preference households have for these foods, and also means that as their incomes rise rural households will switch to field-derived wild vegetables and away from their woodland-derived counterparts.

Further, the positive income elasticities in this table for NTFPs reflect the rather low income levels of Shindi households. Evidence from the regressions suggests that at higher income levels these elasticities would turn negative, so that at higher incomes these NTFPs would actually be inferior goods.

While this table draws on the wider literature on NTFPs, most of the quoted elasticities refer to firewood, as this has been the main quantitative focus of the literature. Three studies of firewood demands conclude that income elasticities turn negative as incomes rise. Further, the other elasticities cited are all either low and positive, or negative. This implies that as households get richer, the absolute quantities of NTFPs that they wish to consume will decrease until at some point NTFP demands will fall to zero. This evidence from formally estimated income elasticities is also backed up by classic observations of consumption shifts that occur as rural households become richer. For example, one immediate shift is from natural construction materials (earth bricks, poles, thatching grass) towards higher-quality purchased substitutes (breeze blocks, zinc roofing). Likewise, richer households choose to consume more purchased foods and beverages and less wild foods and locally produced alcohols. As the income elasticities suggested, richer households also prefer to switch away from firewood towards other sources of energy and lighting.

There are good economic reasons why NTFPs are regarded as inferior goods by richer households. Their production is variable, the quality of the goods cannot be controlled and other substitutes have been bred or produced with higher product quality in mind. Thus more affluent households shift to better substitutes when they can afford to do so. However, the key point is that the inferiority of NTFP goods creates a real problem for NTFP commercialization projects. Dependence on, and use of, NTFPs are linked to poverty and to market failure rather than to household choice: the current prevalence of NTFP use by rural households is a result of their low incomes rather than the attraction of NTFP products themselves. A dramatic illustration of this is given by Hegde *et al.* (1996), who argue that the very high rates of NTFP utilization of their sample households are caused by the fact that these households earn barely more than the Indian minimum wage, and present empirical evidence that as households get richer so their NTFP use declines. So where NTFPs are inferior, attempts to commercialize them will founder on a lack of long-term product demand.

High Transactions Costs of Trading

In our discussion of the stylized features of rural areas, we stressed the role that economic remoteness played in underpinning low rural incomes and in causing the failure of formal product markets. This is also a major cause of the low monetization of rural households' activities, since trading in formal markets outside the rural area is so often uneconomic. These transactions costs will affect NTFP commercialization projects just as much as they would affect any other project aimed at commercializing rural produce. The high unit costs of transportation imply that NTFPs which have a low unit value relative to their weight will have low if not negative margins in the marketplace.

This may be the case for a number of NTFPs. On account of the open-access nature of the resource rules surrounding forests and woodlands in developing countries and the fact that many are inferior goods, NTFP prices tend to be fairly low. Some NTFPs will have high weight-to-price ratios, especially those made of wood. For example, it is noticeable that despite the reasonable prices being attained for carvings on the major roads of Zimbabwe, production activity is contained within a reasonably narrow area, close enough to the road to make transporting the wood inputs an economic proposition. High trading costs have also been implicated in the lack of NTFP commercialization in Latin America. Of course, the converse of this point is the effect that road building has on reducing trading costs and thereby stimulating the marketization of rural produce generally. The impact of the Belem-Brasilia and the Transama-zon highways on economic conditions in the Amazon basin has been widely noted. However, assuming that NTFP projects take the physical infrastructure as given, the impact of transactions costs on the profitability of NTFP commercialization will remain a serious constraint for remoter rural areas for a number of NTFPs.

Storage Problems

Another feature of rural areas we noted was the generally low levels of physical capital. This is true both at the household level and at the community level. Rural areas not only lack household-based productive capital but also the area-wide capital stock, such as power supplies and communications, on which the productivity of household-based enterprises depends. One key problem that this raises is a lack of appropriate storage facilities, such as an absence of refrigeration and refrigerated transport. Indeed, storage of goods is often a major difficulty for rural households. Even for longer-lasting foods such as grains, losses are regularly incurred thanks to mould, rotting and crop pests despite the careful construction of granaries and the use of locally available goods to line and protect these granaries.

Problems with storage could be a major constraint to the commercialization of NTFPs where these goods are perishable. A substantial fraction of the NTFPs examined under the 'Hidden Harvest' programmes are exactly these types of goods: fruits, fruit-based beverages, wild meats and even wild fish. All these would require reasonably capital-intensive storage if commercialization projects were to succeed. Appropriate storage provides two major economic functions, the absence of which would cause problems to emerge. The first function is that of product preservation. Given that the commercialization of NTFPs requires transportation of products to new markets, an absence of adequate storage would result in high losses of products due to spoilage (rotting, bruising in transit, excess fermentation of liquids, etc.). The other function that storage can play is that of interseasonal supply smoothing. Without it, products can only be supplied to the market during the season of their production, resulting in glutted markets, lower product prices and unsmoothed income streams. Thus seasonality and perishability lower the economic returns of NTFPs to rural communities, thereby making these activities less attractive.

Production Risk

Rural households face considerable levels of risk, and one reason for the preservation of woodlands for NTFPs is as a risk-spreading insurance mechanism against severe shocks to the other economic activities of the household. However, in the medium to long term risk-averse rural households will wish not just to spread risk but to reduce it. If NTFP activities themselves still involve considerable production risks, then such activities will be less attractive to rural households than other, more certain income flows. This will be a consideration where the supply of NTFPs is dependent on climatic conditions, such as some of the foods, or where NTFPs are prone to disease.

Open Access and the Costs of Privatization

A further concern derives from the fact that many forests and woodlands in developing countries are held under communal tenure, with usufruct rights to NTFPs very often being effectively open access. If NTFPs were to be commercialized, then one immediate concern would relate to the degree of exploitation of the resource. Raising the value of an NTFP will naturally increase supply pressures on the resource. Where NTFP use is nondestructive to the resource stock, this is not necessarily a problem. However, where NTFP use can diminish the resource stock, then the commercialization of the resource can have serious implications for objectives of environmental conservation. Indeed, there are a number of cases where increases in resource pressure consequent on increased demand for NTFP goods has had serious environmental consequences. One such case is the impact of basket-making on an indigenous tree in Botswana. The main source of leaves for baskets is *Hyphaene petersiana,* which stands up very well to intense harvesting pressure. However, the main source of dye for these baskets is the bark of *Berchemia discolor,* and harvesting the bark can harm the tree if done too frequently. The increase in demand for Botswana baskets has increased substantially the demand for dye inputs, and this in turn has led to the death of many *B. discolor* in basket-making areas. Similarly, the high demand for *Warbergia salutaris* as a source of medicines in Zimbabwe has led to the almost complete extinction of this species within the country. There are numerous cases of this dynamic in the Amazonian rain forests, where palm heart production has resulted in the near elimination of *Euterpe edulis* and *Euterpe oleracea*; the cutting of palms for fruits in Peru has rapidly depleted the numbers of *Mauritia flexuosa, Mauritellia peruviana* and *Jessenia bataua*; and the felling of trees for rosewood oil production had led to the virtual extinction of *Aniba rosaeodora* and *Aniba ducke* in economically accessible parts of Brazil. Commercialization of a destructive NTFP use when held under open access can thus have severe environmental consequences for the species in question (though not necessarily for the forest as a whole).

A longer-term concern relates to the question of who gains from NTFP commercialization. That the existence of communal tenure over woodlands is related to the high costs of exclusion, monitoring and enforcement relative to the benefits to be gained from private woodland control. In other words, the tenurial arrangements over woodlands are related to the ratio of costs and benefits faced by rural households and communities. If as a result of NTFP projects the per-hectare returns to woodland rise, then this ratio changes and economic incentives will arise to change tenurial arrangements. In particular, privatization may emerge as a cost-effective option. However, this is only likely to be cost-effective for households that are already more affluent than others, since only these will be able to bear the private costs of resource control. Two groups stand out here. The first is local élites. There is evidence that as open-access resources rise in value, these resources become co-opted by local élites. The classic study of this phenomenon is Ensminger's (1990) analysis of resource privatization in rural Kenya. In response to the lowering of trading costs via cheaper transportation costs, communal grazing lands were privatized with the consent of rural elders who themselves benefited from this process. However, this process of élite partitioning of commons resources in response to changing economic conditions has also been found in India (Jodha 1986) and Brazil. The second group is outsiders. Town-dwellers or traders/middlemen with superior technology and access to capital may gain control of either the NTFP resources or the NTFP goods' trade and thereby arrogate to themselves the economic returns to these goods. In either case, commercialization-induced changes in local tenure could not only stymie the expected socioeconomic gains from NTFP products to the generality of rural households, but also by triggering the privatization of woodlands and other common resources may leave the rural poor worse off by removing from them a vital consumption back-stop and a source of insurance goods.

Incentives for Domestication and Technical Substitution

We have discussed the many reasons above why NTFPs are currently often low-value goods, such as missing markets, the prevalence of local trading, low local demand and the open-access nature of NTFP resources. One implication of these low values is that neither rural households nor outside agencies currently have an incentive to invest in any sort of resource management or product improvement, since the costs of such actions would far outweigh the benefits. However, this would change if NTFPs were to become higher-value goods.

There are two processes to mention here: domestication and the development of technical substitutes. It is globally true that when natural resources become valuable, there is pressure to intensify production and to domesticate the resource. Take as an example a wild fruit. While prices for the fruit remain low, it is unprofitable for households to do anything other than collect the fruit from the wild. However, as fruit prices rise, at some threshold value it becomes economic for the household (or some other agency) to incur the time and resource costs of establishing the resource under their own control, through planting and tending the relevant species.

Furthermore, domestication also allows resource managers to improve product quality through the development of a constant, uniform or standardized product, thereby maintaining or even increasing product prices. This dynamic is the general story of the development of agricultural products and there is no reason why NTFPs should be any different. Indeed, there are ample examples of cases where rising NTFP values have led to precisely this response. Perhaps the most famous is that of Brazilian rubber *(Hevea brasiliensis)* collection from natural forests. As rubber prices rose in the early part of the century, so it became economic to establish rubber plantations in South-East Asia; more recently, rubber plantations have been established in Brazil as well and these now supply 60% of Brazil's market. Another classic NTFP affected in this way is brazil nuts. In response to rising nut prices, several thousand hectares of *Berthol-letia excelsa* plantations have now been established within Brazil itself. So neither rubber nor brazil nuts provide a solid foundation for NTFP commercialization in the Amazon in the long run. Indeed, it is difficult to point to a wild resource which, once valuable, has remained undomesticated (an extensive list of other Amazonian NTFPs that have been domesticated is given by Homma 1992). Beyond a certain point, then, the commercialization of NTFPs will almost certainly lead to the product being removed from the forest and being placed on the farm instead.

Additionally, if NTFPs become high-value goods, there is an economic incentive to invest in the search for cheaper synthetic substitutes, if this is technically feasible. Once such substitutes are developed, the market for NTFPs often collapses - the 'bust' following the 'boom'. Richards (1993) discusses many such examples of substitute-induced market collapses in commercialized Amazonian NTFPs. At various times these have included dyewood *(Caesalpinia echi-nata),* vegetable ivory *(Phytelephus macrocarpa),* balata *(Manilkara bidentada),* leche caspi *(Couma macrocarpa),* barbasco *(Lonchocarpus* spp.), babaçu oil *(Orbignya phalerata),* cumaru nuts *(Dipteryx odorata),* sorva latex *(Couma utilis* and *C. rigida)* and of course rubber. Such technical substitutions are problems for the entire goal of NTFP-based economic development, since they raise the spectre of entire markets disappearing just as rural households have come to depend on the NTFP resource for a high proportion of their livelihoods.

We should not therefore underestimate the problems associated with the use of NTFPs for the economic development of rural areas. Although rural households currently use NTFPs quite extensively, this is a consequence of the low incomes, capital scarcity, missing markets, riskiness in production and tenure conditions that characterize rural areas. It is these that make it economically rational for households currently to use NTFPs, as these are low value, depend on unskilled labour for their collection and processing, and offset to a degree the riskiness of other production activities.

However, this is no guarantee of success when it comes to the commercialization of NTFPs: (i) there is little evidence that NTFP dependence in the long run is desired at all by rural households; (ii) there are economic problems associated with the costs of both trading and storing NTFPs; (iii) NTFPs still have levels of production risk that may make them unattractive; (iv) the commercialization of NTFPs is likely to lead to the elimination of some species; (v) NTFP development may also undermine the communal tenure system, thereby removing the wider benefits of NTFP use from poorer households; and (vi) NTFP projects aimed at raising product prices will have to face the twin pressures of domestication and technical substitution that have affected almost all other wild products. In the face of these problems, NTFP extraction-based rural economies are likely to be highly unstable, thereby undermining the goal of improved rural household welfare.

7

Forest: An Area of Density of Trees

The forest is a complex ecosystem consisting mainly of trees that buffer the earth and support a myriad of life forms. The trees help create a special environment which, in turn, affects the kinds of animals and plants that can exist in the forest. Trees are an important component of the environment. They clean the air, cool it on hot days, conserve heat at night, and act as excellent sound absorbers.

Plants provide a protective canopy that lessens the impact of raindrops on the soil, thereby reducing soil erosion. The layer of leaves that fall around the tree prevents runoff and allows the water to percolate into the soil. Roots help to hold the soil in place. Dead plants decompose to form humus, organic matter that holds the water and provides nutrients to the soil. Plants provide habitat to different types of organisms. Birds build their nests on the branches of trees, animals and birds live in the hollows, insects and other organisms live in various parts of the plant. They produce large quantities of oxygen and take in carbon dioxide. Transpiration from the forests affects the relative humidity and precipitation in a place.

The FAO (Food and Agriculture Organization) has defined forest as land with tree crown cover (or equivalent stocking level) of more than 10% and area of more than 0.5 hectare. The trees should be able to reach a minimum height of 5 m at maturity in situ. Forests are further subdivided into plantations and natural forests. Natural forests are forests composed mainly of indigenous trees not deliberately planted. Plantations are forest stands established by planting or seeding, or both, in the process of afforestation or reforestation.

Forests can develop wherever the average temperature is greater then 10 °C in the warmest month and rainfall exceeds 200 mm annually. In any area having conditions above this range there exists a variety of tree species grouped into a number of forest types that are determined by the specific conditions of the environment there, including the climate, soil, geology, and biotic activity. Forests can be broadly classified into types such as the taiga (consisting of pines, spruce, etc.), the mixed temperate forests (with both coniferous and deciduous trees), the temperate forests, the sub tropical forests, the tropical forests, and the equatorial rainforests. The six major groups of forest in India are moist tropical, dry tropical, montane sub tropical, montane temperate, sub alpine, and alpine. These are subdivided into 16 major types of forests.

India has a long history of traditional conservation and forest management practices. Under British rule, forest management systems were set in place mainly to exploit forests. Nonetheless, there were some attempts to conserve forests and meet the needs of local communities. The Indian National Forest Policy of 1894 provided the impetus to conserve India's forests wealth with the prime objectives of maintaining environmental stability and meeting the basic needs of the fringe forests user-groups. Consequently, forests were classified into four broad categories, namely forests for preservation of environmental stability, forests for providing timber supplies, forests for minor forest produce, and pasture lands. While the first two categories were declared as reserve forests, the rest were designated as protected forests and managed in the interests of the local communities

Soon after independence, rapid development and progress saw large forest tracts fragmented by roads, canals, and townships. There was an increase in the exploitation of forest wealth. In 1950 the Government of India began the annual festival of tree planting called the Vanamahotsava. Gujarat was the first state to implement it. However, it was only in the 1970s that greater impetus was given to the conservation of India's forests and wildlife. India was one of the first countries in the world to have introduced a social forestry programme to introduce trees in non-forested areas along road sides, canals, and railway lines. A forest is an area with a high density of trees. There are many definitions of a forest, based on the various criteria. These plant communities presently cover approximately 9.4% of the Earth's surface (or 30% of total land area) and function as habitats for organisms, hydrologic flow modulators, and soil conservers, constituting one of the most important aspects of the Earth's biosphere. Historically, "forest" meant an uncultivated area legally set aside for hunting by feudal nobility, and these hunting forests were not necessarily wooded much if at all. However, as hunting forests did often include considerable areas of woodland, the word forest eventually came to mean wooded land more generally. A woodland is ecologically distinct from a forest. The latitudes 10° north and south of the Equator are mostly covered in tropical rainforest and the latitudes between 53°N and 67°N with boreal forest.

CHANGES IN FOREST COVER AND CONDITION

An analysis of changes in the cover and condition of the world's forests requires a differentiation between:

1 the increase in forest cover (by afforestation or natural colonization of trees on non-forest land) or decrease (by deforestation) of forest area; and

2 changes in forest condition, either positive (recovery of degraded stands, stand improvement treatments) or negative (decline, defoliation or dieback, effects of forest fires, degradation through unsustainable exploitation for wood, overgrazing, effects of pests and diseases).

CHANGES IN FOREST COVER

Between 1980 and 1995, the extent of the world's forests decreased by some 180 million ha, an area about the size of Indonesia or Mexico. This represents a global annual loss of 12 million ha, an area equivalent to the size of Greece or Bangladesh. During this 15-year period, developing countries lost nearly 200 million ha of natural forests, mostly through clearing for agriculture (shifting cultivation, other forms of subsistence agriculture, the establishment of cash crop plantations such as oil palm, and ranching). This was only very partially compensated for by the establishment of new forest plantations. Over the same period, forests in the developed world expanded slowly (by some 20 million ha) through afforestation and reforestation, including natural regrowth on land abandoned by agriculture.

Forest plantation and natural regeneration on abandoned agricultural land more than compensated for clearing of forests due to urbanization and infrastructure development in most industrialized countries, outside the former USSR. While the gain of forest cover in developed countries was 20 million ha in the period 1980-95, 9 million ha of this increase occurred in the 5 years from 1990 to 1995.

The situation is quite different in the developing world, where deforestation exceeded net afforestation/reforestation, particularly in the tropical zone. As a whole, the annual rate of deforestation in the developing world between 1990 and 1995 was 0.7%, equivalent to 12.6 million ha, with the highest rate in tropical Asia-Oceania, closely correlated with population and income growth. The lowland forest formations were the most affected by deforestation, although the proportional loss during this period was greater in upland formations (1.1% compared with 0.8% in the lowland formations). There is some evidence that the rate of loss of forest cover in developing countries was slowing towards the end of the 15-year period 1980-95. The annual rate of forest loss in the period 1980-90 was 15.5 million ha compared with 13.7 million ha between 1990 and 1995.

Conversion of Forests to other Land Cover

The Forest Resource Assessment 1990 carried out an assessment of the relative importance of the various factors involved in deforestation at regional and global levels between 1980 and 1990 in the whole tropical belt. Among the most significant outputs of the study were the 'area transition matrices' of the type, which indicate transfers from one land cover class to another over the 3068 million ha of the tropical zone covered by the sample. For instance, the first row shows that 1275.9 million ha of the total area of 1368 million ha of closed forest in 1980 remained as closed forest up to 1990 and also shows the fate of the 92.1 million ha converted to other land cover classes (open forest; long fallow; fragmented forest; shrubs and short fallow; other, *i.e.* non-wooded, land cover; and forestry or woody forest and agricultural tree plantations). All these transfers involved changes of woody biomass attempts to capture by replacing each class along the *y*-axis (average biomass per hectare) and showing the importance of transfers along the *x*-axis.

Changes in Forest Condition

Factors affecting the health and vitality of all types of forests have attracted increasing attention in recent years. Outbreaks of fires have made headline news in developed and developing countries alike, while insect and disease attack and generalized declines in forest condition have caused more localized concern. Insects and disease and forest fires exist under 'natural' systems and conditions; what has caused concern has been their spread in artificial systems and in conditions considerably modified by humankind, resulting in direct economic loss.

Fire arising from natural causes (*e.g.* lightning) has been a major influence and is a driving evolutionary force in several forest ecosystems, such as the pine forests of Central America. Fires are important to the health and maintenance of these ecosystems and the total exclusion of fire from them can lead to the build-up of fuel and ultimately abnormally destructive fires that cause the loss of the ecosystem, as happened with the fire exclusion policy of the National Parks Service of the USA and the disastrous fires of 1988.

However, many fires arising from human activities, whether planned or unplanned, have damaged and destroyed large areas of forests and woodlands. Resource managers face a demanding public often with conflicting needs and incomplete information, leading to the development of new policies that take into account public attitudes towards fire management.

At the global level, an agreement was signed at a meeting in 1990 to initiate the International Decade for Natural Disaster Reduction and the target date of 2000 was set for signatory countries to effectively monitor and manage wildfires in their respective countries. In view of the lack of information on causes, extent and the number of fires at present, and the difficulties in collecting data, this target date is unlikely to be met.

Developed Countries

Total forest area is slowly increasing in the developed world, although the need for protection from fire, insects and disease has continued and some aspects of forest condition give cause for concern. Although the widespread decline and death of certain European forests due to air pollution predicted by many in the 1980s did not occur, deteriorating forest condition has remained a serious concern in Europe and North America. The main causes of the decline of forest condition in Europe have been low soil moisture availability due to drought, and high temperatures. These factors, combined with airborne pollution and the use of non-adapted seed sources in forest plantation development, have predisposed forests to attack by insects and disease and to decline. Forest damage due to air pollution is severe in some parts of central and eastern Europe in particular. Air pollution is also said to have a role in the decline observed in certain species in other industrialized countries, such as sugar maple *(Acer saccharum)* in eastern Canada, the high-elevation forests of red spruce *(Picea)* in eastern USA and *Cryptomeria japonica* stands in Japan.

The exclusion of fire may have caused the decline of some forest ecosystems, such as oak stands in central USA or natural eucalypt forests in Australia where fire is an integral part of the ecology of some plant communities and here it may be used to facilitate their natural regeneration.

Despite an overall increase in the number of fires (by some 40% in Europe, 8% in North America and 120% in the former USSR), wildfires over large areas of forest have become less frequent in the industrialized countries. Further, the average area burned by wildfire decreased in most regions between 1983-90 and 1991-94. There has been a slight reduction in total areas burned in the developed world overall as a result of improved prevention, detection and control systems, although there has been an increase of 15% in the former USSR. In 1990, the average area of forest and other wooded land affected annually by fires in Europe, North America and the former USSR together was about 4.26 million ha or 0.22% of their total area of forest and other wooded land. Pests and diseases remain constant threats, particularly to the semi-natural forests and plantations in the industrialized world. Trees are more likely to be attacked when under stress, whether from abnormal weather conditions such as drought or high temperatures, airborne pollution, uncontrolled movement of germplasm or lack of management. Recent outbreaks of pests and diseases that may have been related to stress include the decline of various species of oaks in many parts of Europe and in central Russia, the recurrent infestation of forests in Poland by the nun moth, the attack on beech by the beech scale in western and central Europe, birch and ash diebacks in north-eastern USA, the littleleaf disease of shortleaf pine in southern USA, and the 'x-disease' of pines in southern California.

FOREST CONDITIONS IN DEVELOPING COUNTRIES

Less attention has been paid to forest condition in developing countries because reduction in forest area has attracted more attention. This reduction in density has often been the result of over-exploitation for timber and fuelwood. Overgrazing and repeated bush fires are other significant causes of decrease in density, especially in the dry tropical and non-tropical zones.

Every year, very large areas of savannah woodland and mixed forest-grassland formations are affected by fires set by herders to provide an early flush of green grass when the rains start, particularly in the dry zones of Africa and South America, although no reliable data are available on their extent.

Forests in the humid tropics have also, at times, been affected by large fires, the most serious in recent years having been those associated with the dry weather conditions arising from the El Niño Southern Oscillation (ENSO). In 1997-98, these caused large-scale destruction of logged and secondary forest in Indonesia (particularly Kalimantan, Sumatra and Irian Jaya), Mexico and Central American countries, as well as leading to smoke pollution over wide areas beyond the borders of these countries. The cause of these fires has been burning for land clearance, but it should be noted that the fires are not necessarily in forest; although described as 'forest fires' they are frequently in grassland or on land that has been cleared of forest. It is *estimated* that of the 2 million ha burned in Indonesia in 1997, 150000-200 000 ha may have been in forest. Other serious fires associated with ENSO effects have included that in East Kalimantan, which burned 3.6 million ha in 1983.

Coniferous forests in the humid tropics have often been affected by fires, although it must be noted that fire is frequently necessary to maintain and regenerate them. In the 1980s, the area of pine forest in Honduras and Nicaragua burned annually amounted to some 65 000ha (or about 3.5% of the total pine forest area of these two countries), and widespread fires in natural and artificial tropical pine forests occur in other countries such as Guatemala, Mexico and Indonesia (northern Sumatra).

In the subtemperate and temperate zones of the developing world, fire is also a permanent threat to forests, particularly when they are no longer used by local people for grazing and other purposes. In the 1980s, the average area of forest and other wooded land burned annually was 140 000 ha in the temperate/ subtemperate zones of South America (including southern Brazil). From 1950 to 1990, fires in China are reported to have affected an average of 890 000 ha annually, the most damaging one having been the 'May 6' fire, which burned some 1.85 million ha in the north-eastern province of Heilongjiang in 1987. In the absence of a global statistical fire database, it is difficult to provide an overall estimate of the annual extent of fires in forests and other wooded lands. A very crude estimate for the temperate/subtemperate and humid tropical zones of the developing world (leaving aside the significant dry tropical zone, for which little reliable information exists) would be of the order of 2 million ha of forest and other wooded land annually during the 1980s. Given the lack of sufficient capacity in fire prevention and control in most developing countries, no significant reduction in wooded areas burned is likely to occur in the near future.

Outbreaks of pests and diseases in developing countries are generally reported for those plantations and planted trees where the impact is most apparent. Introduced pests are usually extremely destructive and several of them have had very damaging effects in recent years in the developing world. Examples of introduced insects affecting plantations and planted trees include the *Leu-caena* psyllid, *Heteropsylla cubana,* which has spread into Asia and the Pacific islands and is now extending across Africa; the cypress aphid, *Cinara cupressi,* which is established in eight eastern and southern African countries and is causing heavy mortality among a number of exotic and indigenous species but especially the important plantation species *Cupressus lusitanica,* threatening its future role in the plantation programmes of these countries; and the European woodwasp, *Sirex noctilio,* which has spread into Argentina, Uruguay and southern Brazil, affecting principally *Pinus taeda* but which may become a threat for the large Chilean plantation estate of *Pinus radiata.*

Strategies to control introduced insects from attacking forest trees include regulatory measures, eradication, integrated pest management, and monitoring of population levels and occurrence of potentially harmful species. Effective monitoring of insect populations will be especially important in view of the large plantation programmes being carried out to produce industrial roundwood, which often rely on one or a few exotic species. Such programmes may be able to afford the investment that is necessary for control measures. Smaller-scale programmes may have to insure against failure due to unexpected attack by insects or diseases by diversifying into several species adapted to the sites and end-uses.

There is also evidence of forest decline due to a combination of biotic and abiotic factors in the developing world, with air pollution likely to play an increasing role in some cases as industrial and transportation infrastructure develops. Examples of such decline include neem *(Azadirachta indica)* in the Sahel, framiré *(Terminalia ivoren-sis)* in Ivory Coast and Ghana, and *Eucalyptus globulus* plantations in Colombia and Peru, while airborne pollution is affecting forests near large cities and industrial areas in China (*e.g. Pinus massoniana* stands near Nanshan). Despite the overall significance of the corresponding damage and losses, surveys of forest decline and diebacks in developing countries remain all too rare.

RECENT ESTIMATES OF GLOBAL FOREST AREA

The following section is based on 1990 baseline figures, updated to 1995, prepared by the Food and Agriculture Organization (FAO) Forest Resource Assessment Programme, reported in and largely drawn from *State of the World's Forests 1997.*

In 1995 forests were estimated to cover 3454 million ha, or 26.6% of the total land area of the world (Greenland and Antarctic excepted). Almost two-thirds of the world's forests were located in seven countries: Russia, Brazil, Canada, USA, China, Indonesia and Zaire; 29 countries had more than half of their land covered by forest, of which 21 were in the tropical belt. However 49 countries, in addition to the many non-forested small island states and territories, had less than 10% of their land covered by forests. Five entire subregions were in this category: North Africa (1.2% of the land area), Near East (1.9%), Temperate Oceania (6.2%), Non-tropical Southern Africa (6.8%) and West Sahelian Africa (7.5%).

The latest information on the distribution of forest and of forest cover change by ecological zones was provided in 1990; the results of the Forest Resources Assessment 2000 were released by FAO in 2000. In 1990, temperate and boreal forests occupied 1.64 billion ha and tropical forests 1.76 billion ha. A breakdown by ecological zone was available only for tropical forests.The figures on tropical forests show that most (88%) are in lowlands; of these, tropical rain forests accounted for 47% of all lowland tropical forests, followed by moist deciduous forest (38%) and dry and very dry formations (15%).

Natural Forests

The forest area estimates described above include undisturbed forests, forests modified by humans through use and management (or 'seminatural' forests) and forests created artificially by humankind (*i.e.* forest plantations) by afforestation or reforestation. (Afforestation is defined as the establishment of a tree crop on an area from which it has always, or for a very long time, been absent. Reforestation is defined as the establishment of a tree crop on forest land.) In most industrialized countries, particularly in continental Europe, forests are being managed in such a way that at management-unit level a continuum exists from low-intensity management, involving natural regeneration, through more intensive methods involving some artificial planting to highly intensive methods with complete planting and cultivation; this makes it difficult to isolate figures for natural forest and plantations. The distinction between natural or seminatural forests and forest plantations can more easily be made for developing countries and some industrialized countries such as New Zealand in which forest plantations have been established using introduced species.

Interest in natural forests, particularly their role in the conservation of biological diversity, has led to efforts to compare forests today with what is thought to be their original character and to give complete protection to areas of forests which have had no, or minimal, human interference. Although there are difficulties in identifying the extent of natural forest, compounded by problems of definition, some information exists that can be used as an indication of broad patterns of natural forests in various regions.

An attempt has been made by the World Wide Fund for Nature (WWF) to quantify the area of forests in western Europe that has been relatively undisturbed by humans or which has retained much of its natural character. Their report distinguishes between 'virgin forest', defined as 'forest ecosystems whose characteristics are determined exclusively by natural location and environmental factors... without human influences present or visible any more', and 'natural and ancient seminatural forests', which 'have not been planted or sown by man for the past two centuries' and 'which continue to have a large number of the natural elements'. The study found that only a small proportion (probably < 1%) of the total forest land in northern and western Europe could be considered as virgin forest, which has arisen since the last glaciation. Almost all was located in Sweden, Finland and Norway, with small areas in Greece, Austria and Switzerland and (according to another author) in France. In eastern Europe, Slovakia and Belarus have considerable areas of virgin forest while Poland and Croatia have small areas. In addition, in northern and western Europe, the WWF report identifies natural and ancient seminatural forests representing 2.1% of the total forest cover (1990) of the 16 countries concerned. There are a further 3 million ha of land in the region in national parks and other protected areas and another 50 000 ha in small forest reserves (mostly for nature conservation and scientific research), whose use is tightly restricted.

The situation in temperate and boreal North America is quite different from that of densely populated Europe and Japan, where use and management of forests for many centuries have left very little of the original forest area untouched. 'Old growth forests', as they are called in North America, still cover extensive areas.

On the lands managed by the US National Park Service alone, old-growth forests covered 1.97 million ha in 1988. Although not a direct measure of the extent of old growth forest, it may be noted that the area of forest included in national parks and other protected areas in North America (USA and Canada) was reported to be nearly 49 million ha in 1990.

The FAO Forest Resource Assessment 1990 did not distinguish between undisturbed and disturbed 'natural' forests in developing countries. However, the Forest Resource Assessment 1980 made estimates of the areas of undisturbed closed forests (primary forests and old secondary forests where there had been no logging for the last 60-80 years) and of closed forests that were included in national parks and other protected areas (thus relatively undisturbed, at least in theory). At that time, these two categories together represented 60% of the total closed forest area in the tropics, a proportion varying from 39% in tropical Asia to 59% in tropical Africa and 69% in tropical America.

These different proportions by region reflected a slower development of large-scale harvesting in tropical America compared with tropical Africa and

Asia, and also the fact that, in tropical America, spontaneous colonization did not follow in the wake of logging as systematically as in the two other regions due to lower population pressure. Although the two categories do not match the concept of 'virgin forests' of Europe and of 'old growth forests' of North America, the sum of the two categories nevertheless gives an indication of the amount of forest disturbance (or management, depending on the point of view of the observer) that existed around 1980 in the humid tropics. Although corresponding estimates for 1990 and 1995 are not available, it is likely that the share of undisturbed forests remains higher in the three tropical regions than in Europe and probably in North America too.

GROUPS OF VASCULAR PLANTS IN FORESTS

Before we go further into the historical development of forests and forest types, it might be well to note briefly the various major groups of vascular plants. This is perhaps advisable because several groups which dominated the forests of the past are now almost extinct, having left behind only a few living remnants which no longer are arborescent. Were it not for their pasts they would be outside the scope of this discussion.

The designations of the major groups of vascular plants may not be familiar to all. Formerly it was customary to divide the vascular plants into two groups, the *Pteridophyta* (reproducing by spores, as in the ferns, horsetails and lycopods) and the *Spermatophyta* (reproducing by seed). Advances in our knowledge have revealed that this simple division is no longer tenable.

Currently seven major groups may be recognized; these appear to have been independent as far back as we have knowledge of them in the fossil record. Fortunately, each of the major groups as now defined still contains living members. Therefore each group name has been derived from the name of a living genus, athough the extinct genera may have been far more numerous. Thus, the *Equisetopsida* are more or less like the modern horsetails *(Equisetum)* and the *Magnoliopsida* (the so-called "flowering plants") are more *"Magnolia*-like" than "fern-" or "*Pteris*-like" *(Pteropsida)* and so on. In the following text, informal usage is sometimes made of the more formal group designations.

The present discussion cannot do more than briefly mention certain matters relative to the development of these groups. Acquaint us with the names of these major groups of vascular plants, their most prominent subgroups, their known extent in geological time and their general prominence in the world of the past as revealed by the fossil record. However, at all times we must remember that the "fossil record" to a great extent is composed of members of lowland floras which were present near estuaries or in descending basins of deposition and whose remains were thereby readily entombed in silts and muds by seasonal floods, or in the acid muck of great swamps then essentially at sea level.

The variety of materials then growing on the much more extensive uplands can be conjectured only from clues in the form of logs and miscellaneous scraps which floated downstream, probably for long distances and so were preserved in the deposits as alien waifs along with the far more abundant remains of entirely different types of the "local" lowland vegetation characteristic of the time.

Psilopsida

Of these major groups of vascular plants, only the psilopsids did not develop arborescent members, perhaps because of an inherent inability to produce either a ring of secondary wood or a soundly functional root system capable of anchoring a heavy superstructure.

Aneurophyton

The exact affinity of this material has not yet been ascertained. Reproductively it was psilopsid in character and so usually has been classified with that group. However, it had considerable secondary growth and a good basal root system and hence became markedly arborescent; and it differed from typical psilopsids in other characters. Some years ago this plant received much publicity under the name of *Eospermatopteris,* the supposition being that it bore seeds. Investigations with newer techniques revealed that the so-called "seeds" found associated with extensive remains of it near Gilboa, in eastern New York State, actually were spore cases — and of another and apparently unrelated plant. With the discovery of the true nature of the trees making up the "Gilboa Forest," it became evident that *Eospermatopteris* was the same plant as the earlier described *Aneurophyton,* which name it now properly bears. As our knowledge of this ancient form increases, it may be found to have played a more important ancestral role than we have so far supposed. On the basis of modern concepts, *Aneurophyton* may yet be moved from its present position in the series to one perhaps between the pteropsids and cycadopsids.

Lycopsida

The four remaining genera of this major group tell us little of the grandeur of the Carboniferous lycopsid forests. The earliest known member of the group, *Baragwanathia,* so simple that sometimes it has been classed as a psilopsid, probably was not ancestral to the entire group; if anything, it gave rise to derived forms from which our modern lycopods have descended. The lepidodendrids, a motley group, had a series of components, only a few of which seem to have come directly from the geologically preceding protolepidodendrids; all these seemingly had an indirect origin out of a common phyletic plexus, perhaps during the Silurian. The lepidodendrids, often with ample secondary growth and special systems of internal supporting tissues, were truly arborescent. Of interest, however, is the fact that both the aerial portions and root systems were dichotomously branched and similar in organization, except for such differences as we would expect between aerial and subaerial structures. Furthermore, the ultimate absorbing "rootlets" were coarse structures, seemingly little more than highly modified leaves. Hints of this old lepidodendrid system of organization still are retained in a few modern lycopsids such as *Isoetes.* The arborescent lycopsids were not engineered to meet the competition of the structurally and physiologically more efficient groups that were to come into prominence later during the Permian period.

Equisetopsida

The hyeniads, a curious group of smallish, almost shrublike plants, appear to be ancestral to the calamites and equisetums (the modern horsetails). The sphenophylls probably were a cognate group, at all times in their history rather weak and sprawling members of the swamp and, perhaps, moist upland communities. The arborescent calamites, abundant in the swamplands of the Carboniferous, had considerable secondary growth, yet in their organization and plant habit they were remarkably like our modern horsetails. These, in turn, may be little more than greatly reduced representatives of a collateral line developed from a basal member of the variable calamitean group no later than the Carboniferous and perhaps even in the Devonian.

Pteropsida

Although arborescent pteropsids were relatively abundant during the Carboniferous and since have been present in the several lines of this group, their arborescent nature always has been anomalous. The actual woody portion of their "trunks" never was large, the trees being primarily supported by a heavy mass of tightly intertwined roots which, originating at the bases of the leaves, grew downward, encasing the much weaker actual stem and thereby holding the plant erect. Nothing really new has happened in the basic architecture of the arborescent pteropsids since the Carboniferous. The development of numerous herbaceous and climbing pteropsids, as well as the derivation of forms suited to desert and aquatic habitats, may be merely noted.

Cycadopsida

The early history of our knowledge of this group is largely one of misunderstanding. For many years men have been astonished at the wealth of fernlike leaf impressions associated with Carboniferous strata. This easily led to the assumption, still maintained in some texts, that the Carboniferous was the "Age of Ferns." Coenopterid ferns were present, but for the most part they were both so uninteresting and so unfernlike as to be passed by. With the discovery that some of these fernlike leaves actually bore seed, a long controversy was started as to whether such forms were seed-bearing ferns, or seed plants with fernlike foliage.

Continued work on the group revealed that some never did bear seeds and also had the internal structure of true pteropsids. The bulk, however, appear to have been seed plants with fernlike leaves. More recent investigations, with improved techniques, have uncovered a large and complex series of these "seed-ferns," or pteridosperms and have also brought to light what, at last, appears to be excellent evidence that the majority were in no way related to the ferns but are close relatives of and in part ancestral to, the cycads. However, as we shall note later, a few of these so-called pteridosperms actually may be in the magnoliopsid line.

The cycadopsid pteridosperms were a varied lot. A number of them were arborescent, with abundant secondary growth in the roots, trunks and branches; some may have been of considerable stature. Others were shrubby, quite a few are known to have been high-climbing and anatomically complex woody lianas and there are hints that a number of herbaceous pteridosperms also were present. Long before they passed out of existence, the pteridosperms gave rise to both the cycads and cycadeoids.

Recent investigations of their anatomical structure indicate that various of the pteridosperms were capable of withstanding periods of drought. It may have been such forms which gave rise to the characteristically more xerophytic, arborescent cycads. Another derived group, the cycadeoids, differing from the cycads in various ways, seems to have been peculiarly adapted to semidesert conditions. These thick stemmed (but always lowgrowing) cycadeoids also evolved a "flower," but in its structure it was a cycadopsid flower and not one which could have given rise to the magnoliopsid flower. The few remaining species of the tropical, arborescent gnetums also appear to have been derived from some ancient group of pteridosperms.

Pinopsida: As here defined, the pinopsids encompass four collective subgroups: the extinct cordaites, the ginkgos, the "conifers" (some of which do not bear cones) and the taxads. The one living species of *Ginkgo* retains a primitive leaf structure and an even more primitive type of reproductive apparatus with swimming sperm; this latter feature also is characteristic of the cordaites and of all known "conifers" until after the Permian.

The taxads appear always to have stood somewhat apart from the other pinopsids, the wood structure of the group being the main clue to its basic affinity. To delve further into the intragroup relationships of the pinopsids in this place would open a controversy that is yet far from being resolved. But our uncertainty as to the exact relationships of the various pinopsid subgroups is a matter of insignificance in the face of the mystery of the origin of the group as a whole. In the upper reaches of Devonian strata, in deposits of estuarine and offshore nature, numerous trunk segments occur.

Microscopical examination of these, when well preserved, reveals that they constitute a series of pinopsid forms, some already with a highly evolved wood structure. All we can say is that these numerous trunk segments prove that the pinopsids are a very ancient group and that already in the Devonian the uplands of the world were clothed with pinopsid forests. From what we know of them, various of these ancient trees must have appeared much like modern araucarians.

Magnoliopsida: This group traditionally has been called the "angiosperms" (with covered seed), to distinguish it from the "gymnosperms" (with naked seed). A recent division of the so-called gymnosperms into *Cycadopsida* and *Pinopsida,* made necessary by their obviously different lines of descent, left the "angiosperms" standing unique without a name comparable to those of other groups. Therefore a widespread, well-known and relatively primitive genus, *Magnolia,* has been chosen to loan its name to this major group. The dicotyledonous hardwoods as well as the monocotyledonous palms constitute such important factors in our forest economy and are so familiar to all of us, that any discussion of their intragroup relationships in this place would be superfluous.

The often repeated concept that the magnoliopsids "came into being and underwent a great burst of evolution in the Cretaceous" is erroneous. Originally an upland group and therefore but rarely preserved in the earlier fossil record, the magnoliopsids found no opportunity until the middle portion of the Cretaceous to migrate in sufficiently large numbers into the lowlands to become commonly and widely preserved. When the group becomes at all commonly represented, we discover that already essentially all its modern arborescent families had been evolved and many of its modern genera also already had come into being, often with a series of species surprisingly similar to those of today. This indicates only that the magnoliopsids are not of recent origin, that they are no more "modern" than other groups of vascular plants. But the origin of the magnoliopsids is another matter.

Among the so-called pteridosperms may be found a series of forms which, even by the most conservative criteria, would be classed as aberrant for the group. In many ways these fit into our concepts of what, on purely theoretical grounds, we should expect to be forms ancestral to the magnoliopsids. But these aberrant pteridosperms were "gymnospermous" in their reproduction, whereas the magnoliopsids are typically "angiospermous" — and so the two have been kept separate by most workers.

But before we thrust these seemingly insignificant aberrant pteridosperms, with their seeds borne on modified leaves, back into the cycadopsids because of the gymnospermous nature of their reproductive apparatus, let us recall that there exists a series of magnoliopsid genera, admittedly primitive in the great magnolialean-annonalean complex of tropical, arborescent forms, which, even today, open their sometimes remarkably leaflike carpels at flowering time, thus exposing the naked ovules (the future seeds) in a typically "gymnospermous" form of pollination. We now suspect, also, that some of the wood fragments of the Devonian, currently assigned in a general way to the pteridosperms, on more careful examination may yet be found to be the wood of primitive gymnospermous members of the largely angiospermous magnoliopsids. A traditional passion for large and sometimes not too precisely defined terms in our so-called "scientific language" should not be a semantic block keeping us from an understanding of the true origins and developments of our great groups of vascular plants.

THE FIRST FORESTS

Direct fossil evidence of a vascular land flora appears for the first time in Silurian strata through a scattering of psilopsid forms and the primitive lycopsid *Baragwanathia*. Primitive as these forms appear to be in the light of what came after, they seem highly advanced when we consider the differences between them and any hypothetical, vascularized alga-like form from which they might have descended. There is considerable evidence that the first steps in the evolution of a group of organisms is a slow and halting process, accompanied by many excursions into biological blind alleys and beset with numerous failures. Therefore we can but conclude, on the basis of the known forms of the Silurian, that a vascular land flora of some sort already existed in the Ordovician and gave rise to that of the Silurian.

Unfortunately for us, the Silurian was a time of continental uplift and therefore one with few opportunities for the preservation of representative examples of what must have been an interesting succession of evolutionary experiments in the perfection of successful land-dwelling vascular forms. The Silurian culminated in a series of almost world-wide disturbances, accompanied by the building of great mountain ranges, which probably played much the same role that mountains always have in the development of vegetation types subsequently to become common and dominant in the lowlands.

The seemingly abrupt appearance, in the lowland deposits of the Devonian, of various forms of all major groups of vascular plants except the *Magnoliopsida* indicates that they must have been present and undergoing an active evolution on the uplands during Silurian time. And, as noted earlier, there is a possibility that certain wood fragments of the Devonian, now assigned to the cycadopsid pteridosperms, actually may be of materials in the magnoliopsid line; if this proves to be so, the magnoliopsids would then have a recorded lineage just as ancient as that of any of the vascular groups which gave rise to arborescent forms.

By Middle Devonian time the lands of the world were dominated by a series of forest types. Aneurophyton forests, mixed with gaunt-branched protolepidodendrids, were widespread in the lowlands; moist and swampy areas bore abundant psilopsids, hyeniads (the earliest known of the equisetopsids) and a variety of early coenopterid ferns.

The abundant segments of waterworn trunk and other wood fragments which began to appear in the deposits indicate the presence on the uplands of extensive pinopsid forests, probably with an understory of somewhat smaller pteridospermous trees and an undergrowth of lesser sorts. Except for the dwindling and loss of a few less specialized groups, the forests of Upper Devonian time were largely a preview of what was to follow.

FORESTS OF THE CARBONIFEROUS

The Carboniferous saw the extensive development of lowland forest types, as we should expect of a period when the continents were generally at a low level. Again, however, we must presume a certain amount of upland, even with mountains of possibly considerable height, for there are far too many fragments of trunk and branch wood, with no associated leaves or other structures with which they can be related, to be accounted for in any other way.

Furthermore, where else but from adjacent mountains and extensive uplands could the sediments producing the abundant sandstones and shales of the time have been derived? But it is to the lowlands, with their rich fossil record, that we turn our immediate attention. It has usually been thought that the extensive Carboniferous lowlands were characterized by extremely heavy precipitation and there is no doubt that swamps were widespread. What sort of "swamps" were they? The answer is to be found in the trees that inhabited them.

To live and function actively every cell must have a relatively abundant supply of oxygen and also be able to get rid of any excess of carbon dioxide produced during certain physiological reactions. Therefore, when we examine the plants of constantly wet areas, we find that those portions where the aeration is poor develop a sort of internal tissue known as aerenchyma.

This specialized tissue consists of a series of ramifying ducts, channels, or connecting loose-celled structures whereby the living cells of the submerged portions can exchange air with the parts in normal atmosphere. There is little point in arguing, as some have done, that "plants were different then." Protoplasm was protoplasm, just as it is today and there is absolutely no evidence that it had any different requirements or tolerances. The forms of the plants have, perhaps, changed somewhat (fundamentally very little), but the basic physiology seemingly has not. Therefore, we can examine the roots of these groups of Carboniferous plants and tell with considerable precision whether or not the soil in which they grew was saturated with water.

Of the roots of Carboniferous plants that have been examined in any detail, only the following are known to have possessed aerenchyma: the equisetopsids in general (the sphenophylls and arborescent calamites), the marattiaceous tree fern *Psaronius* and perhaps a few, but by no means all, of the arborescent lepidodendrids. The great bulk of the lepidodendrids and all the other arborescent forms of the time, so far as is known, were devoid of it, indicating that they did not inhabit areas which were flooded and waterlogged for considerable periods. But how, then, can we explain the great deposits of coal in the Carboniferous, if they were not laid down in vast, reeking swamps? Here again the Coal Measure strata tell the story.

These abundant plant remains accumulated in slowly descending basins. Near the estuaries, which often extended well into the continental interiors, in actual depressions inland and along streams, the equisetopsids, *Psaronius* and a few of the many species of lepidodendrids, with their specialized root aerenchyma, were abundant. With an equable and completely frost-free climate extending at least to latitudes 60° N and S, there would be ample opportunity for rampant growth over a large portion of the land surface if soil moisture were reasonably available.

In the slowly descending basins characteristic of large areas of the land surface at various periods of the Carboniferous the soil moisture was certainly ample at all times of the year. However, this does not mean that the rainfall was necessarily excessive or the soil saturated with water, except locally. With the generally low relief of these basins would have been associated a complex pattern of saturated, poorly aerated soils, intermingled with soils that were moist but had sufficient aeration for the growth of plants without a specialized aerenchyma in their roots. So intricate was the pattern and so dense the vegetation that as the trees fell, to be preserved in the deposits, they fell in such a way as to make it seem at first glance as if they had all grown in exactly the same habitat. The same intricate pattern in the distribution of different species is easily discerned in our modern swamp forests, where we find some species capable of persisting with their roots submerged, while others, growing so close as to have their branches touching, may be seen to occupy slight rises on which the soil, although constantly moist, is much better aerated. These latter species also are likely to be found wherever the soil moisture is abundant outside the swampy zone, sometimes even on uplands.

Although moisture was reasonably constant in the lowland soils, the humidity of the air was not excessive. Careful study reveals that the great majority of the Carboniferous plants had leaves which in some manner resisted water loss through transpiration. This applies to the arborescent lepidodendrids, the pteridosperms and the cordaites — the last a group of lowland pinopsid trees then in great abundance, which sometimes formed. dense stands. Even the microphyllous equisetopsid trees of the time, growing in their soggy environments, indicate that the atmosphere was by no means saturated with water vapour.

For our purpose, however, it is perhaps not overly important what the conditions were under which these lowland forests grew, except insofar as they shed light upon the worldwide atmospheric and climatic conditions of the time and thus give clues as to what was probably happening on the even drier uplands, where the ancestors of our contemporary trees already were undergoing their primary development.

Actually, we have little first-hand evidence regarding the upland forests of the Carboniferous, apart from what is revealed by the relatively few fragments of trunk and branch wood that floated into the lowlands, there to be preserved in the deposits.

While these give us valid hints, it is from the groups that migrated into the more equable lowlands during the severe climate of the Permian that we can glean a more satisfactory picture of the upland forests of the Carboniferous. Just as the wood fragments indicate, these migrants consisted mostly of a series of pinopsid forest types, with forerunners of the true "conifers" dominant among them. Often made up of giant trees (specimens with trunks up to six feet in diameter are known), these pinopsid forests probably had an understory of pteridospermous trees, shrubs and vines (with perhaps a few herbaceous types) and there is every likelihood that ferns of diverse sorts were abundant.

The subhumid areas near the 30° parallels probably were not devoid of vegetation. The tough-leafed pteridosperms of the time were already giving rise to the cycads and cycadeoids, both admirably fitted to less humid — even xeric — conditions. Therefore, it is probable that the less severe parts of these subhumid areas supported a variety of pteridosperms suited to such conditions. And a few scattered fossils indicate that ancestral elements of certain lines, leading to those of the modern pinopsids better suited to dry areas, already were present. Cordaitean forests dominated the moist lowlands; ancestral "conifers" were beginning to take over the uplands.

If any forests grew in the cool-temperate regions and on the slopes of any higher mountains of the Carboniferous period, their nature is largely a matter of conjecture. If, however, the requirements necessary in the present are a mirror of those of the past, there is every reason to believe that abundant forests were then present in the Arctic and Antarctic regions, unless they were limited in some manner by the lack of light during the long polar winter; certainly the temperatures were no bar to the development of dense forest stands.

If so large a number of pinopsids can today stand temperatures far below freezing, surely the large and variable group then extant could have included forms capable of persisting in the polar regions in the less rigorous temperate climates of the time. Their apparent abundance on upland habitats at middle latitudes during the period is an almost certain indication that the pinopsids also included a wide series of types adapted to temperate climates. And since resistance to frost in plants is physiologically closely akin to resistance to drought, there is considerable reason to believe that the poleward temperate regions of the time also supported at least a scattered assortment of pteridospermous types.

Although there is little or no evidence that the vast lowland forests of the Carboniferous contributed anything to the development of modern arborescent types, they played a most important role in relation to the use and ultimate preservation of our present-day forests. Before the widespread use of coal, wood was almost the only source of fuel. In viewing our dwindling forest resources, we may be thankful that the lowland forests of Carboniferous time, entombed in the strata of the Coal Measures, have relieved our modern forests of much of this great and potentially destructive burden.

THE GREAT CHANGE IN THE PERMIAN

The Permian opened with a climate very similar to that of the latter parts of the Carboniferous and with the same general distribution of forest types. This pattern was soon disorganized by what may have been the greatest period of glaciation that the world has ever experienced.

By comparison, it dwarfs any of the glacial cycles of the Pleistocene, which terminated the Tertiary. In the northern hemisphere evidences of Permian glaciation are known from Alaska, Massachusetts, England, France, Germany, the Ural Mountains and India. In the southern hemisphere all of what is now Australia was covered and extensive glacial deposits of the same age are reported from Madagascar, various parts of southern and southwestern Africa and South America, notably Brazil, Argentina and Bolivia. Nor is there any hint that these deposits were derived from local mountain glaciers; they were of a continental type.

The evidence of widespread glaciation during the Permian in the southern hemisphere, coupled with that of an amazing glaciation in what is now tropical India in the northern, did as much as anything to spark the far-from-settled controversy concerning the possibility of continental "drift." If, as postulated by the proponents of the theory, these glacial deposits could have been laid down in polar regions and then later transported into what ordinarily have been thought of as the temperate and tropical regions by a lateral shifting of the continental masses, there would be no need to disturb the "normal" climatic belts of the world.

The possibility of at least a certain amount of continental displacement need not be ruled out. But, the story told by the forest types of the time indicates that a vast world-wide change in climates actually did take place: those groups which, since the Devonian, had been limited chiefly to frost-free regions and apparently fixed in their evolutionary potentialities were essentially wiped off the face of the earth. The climatic situation during the height of the Permian glaciation. The completely frost-free zone may have dwindled to a narrow band in the equatorial regions, thus cutting down a multitude of tropical genera and species. The continental uplifts would have drained the great Carboniferous swamplands, thereby eliminating other arborescent types with which the world had been so long familiar and interior plateau deserts with severe climates would have been extensive. It was a time of crisis for all the old forms of life — animal and plant and many of them failed to meet the challenge.

Since there is little point in dwelling on what disappeared, let us turn to that which survived. Actually, the frost-free areas may have been rather more extensive than some students have supposed, for a few of the Carboniferous lepidodendrids, calamites, cordaites and pteridosperms passed through the perilous times of the Permian and lingered on into the Triassic. Furthermore, the fossils tell us that the land unoccupied by glaciers was never barren of vegetation except in the worst of the and areas and in the polar regions; there were vascular plants of one sort or another right up to the ice fronts. And the rapid development of tropical lowland forests in the Triassic also points to the presence, in the Permian, of considerable areas of at least warm-temperate climate.

We leave this most interesting and critical period in the world's vegetational history with only one last glance. Various modern types of fern had become more evident. The cycads and cycadeoids were now common and venturing into the drier habitats then widespread. The ginkgos, the true "conifers," and perhaps even the earliest of the taxads were abundant, not only in the lowlands, but also on those uplands where the climates were not too severe.

The pteridosperms, although on the wane, still displayed a wide variety of forms. As previously noted, among these were groups which, for want of a better term, might be collectively called the "aberrant pteridosperms" — plants often with entire margined leaves and a reticulate venation. The male reproductive cells, or sperm, still were motile and for this reason they have traditionally been classed with the pteridosperms, although recognized as distinct orders and families. Are we here getting our first glimpse of the real ancestors of the magnoliopsids? If these considerably varied "aberrant pteridosperms" actually were an independent complex of forms out of which the magnoliopsids were derived, then the Permian marks the really important point of change in the floras of the world, a closing out of the old forms and a bringing into prominence of the basic members of all the important modern arborescent groups.

THE GREAT PINOPSID EXPERIMENT OF THE TRIASSIC AND JURASSIC

Contemporaneously with the dwindling of the Permian ice fields and the return to a more equable climate in the Triassic, something approaching an ecological vacuum came into being, for the old forms dominant during the Carboniferous had essentially disappeared. The disturbances of the Permian had given rise to a large series of upland habitats, which were exploited by the remaining plant groups of the time.

With the newly available, although still relatively small, lowland areas freed from competition by the old types, the newer types filled the many vacant ecological niches of these regions. Being in the ascendency and already the dominant upland forest type, the pinopsids began to develop an extremely divergent series of derived lowland forms. Many of the Permian types held on into the Triassic, but in the expanding frost-free lowlands broad-leafed pinopsids began to be evident, suited to such environments. Certain modern pinopsids, such as the broad-leafed podocarps, appear to be collateral descendents from just such types. And it seems likely that a few, such as the modern *Phyllocladus,* abandoned leaves and produced broad leaflike structures by greatly flattening and expanding their lateral twigs. In the temperate lowlands further varieties of leaf forms showed themselves, some suited to humid and others to drier conditions.

In Triassic times deserts, residual from those of the Permian, or produced in the rain shadows of far-flung mountains, probably existed and the widespread occurrence of numerous cycad and cycadeoid types suggests large subhumid areas. But the giant pinopsid trunks, abundant in the Petrified Forest of Arizona and elsewhere in Triassic formations, testify to extensive forests and the opinion expressed by some that the Triassic world was largely dominated by true desert climates is probably erroneous. In the earlier stages of the Triassic the land masses were still elevated and so afforded relatively few places where adequate deposits could be formed; therefore, we know little of what must have been the highly variant forest types of the time, save what can be deduced by observing the situation during the transition to the Jurassic.

A further lowering of the land masses in the Jurassic led to considerable amelioration of the climates of the world. Since it also gave opportunity for the formation of more extensive deposits, we know considerably more of the forest types of the Jurassic than we do of those of the Triassic. The cycads and cycadeoids were still abundant, indicating the persistence of subhumid areas of considerable extent. Along with the many old and now extinct groups of pinopsids, modern genera began to appear. About this time the ginkgos perfected a functional abscission layer in their leaf bases, which permitted their leaves, instead of hanging until tattered and unrecognizable, to be blown about as they fell off the trees each autumn and so to settle in depressions and become preserved. These characteristic and abundant leaf impressions permit us to ascertain that the group was widespread and common in the northern hemisphere, probably with a goodly number of species.

The great bulk of the fossils has led to the conclusion that the forests of the Triassic and Jurassic were dominated by pinopsid types. But let us again remember that, although the continental masses apparently were being slowly lowered, there still were broad uplands and considerable mountainous territory, whose covering vegetation is necessarily almost unknown. As Axelrod has pointed out, the earliest materials which can be placed among the flowering plants (magnoliopsids) with any degree of reliability have been found in Triassic strata. Materials about which there can be no argument are of increasing frequency in the Jurassic. Furthermore, by the Jurassic even these fragmentary remains indicate substantial diversity of form, with evidence that both monocotyledonous and dicotyledonous types already were in evidence. There is only one conclusion: beginning with the Triassic and more abundantly in the Jurassic, the upland and mountainous regions, as distinguished from the more widespread areas of deposition, bore an increasingly diverse assortment of surprisingly modern magnoliopsid forest types.

By the end of the Jurassic the great experiment in diversification by the pinopsids was almost over. They had filled the numerous and often special ecological niches left vacant by the passing of the lowland forest types of the Carboniferous and Permian. They had not only achieved world-wide dominance, but had produced a diversity of kinds which they were never to display again. Of the 55 genera of pinopsids still living, 47 may be classed as relics in view of their restricted or highly disjunct present distributions, or else because their distributions today are much smaller than were those shown by their remains in Cretaceous and Tertiary deposits.

THE SORTING OF THE FLORAS IN THE CRETACEOUS

Complex and varied as they were, the pinopsids between the Permian and the Cretaceous are known from such scattered localities that it is difficult to determine and compare the distributions of the various families and genera. It becomes obvious, however, that even before the Cretaceous there was a strong divergence between genera, some being predominately of the southern hemisphere and others characteristic of the northern. This separation has since been greatly intensified. Today only a few pinopsid genera have occasional species that cross the equator and these are mostly along transequatorial mountain ranges, where they may be recently adventive.

The more critically we study our flowering plants (magnoliopsids), the more it becomes evident that the groups we are prone to think of as characteristic of the temperate regions actually have their more primitive living members in the frost-free parts of the world or have been patently derived from other groups essentially tropical in composition. In view of what we know of their distribution in the Jurassic, in conjunction with the foregoing situation, we are increasingly led to the conclusion that the magnoliopsids underwent their primary evolution, not in the lowlands, but on the uplands of frost-free regions. Since there is reason to suppose that the magnoliopsids were actually present during the earlier phases of the Triassic, this primary evolution would have taken place roughly between latitudes 30° N and S, the probable limits of the frost-free conditions of the time.

Forms suited to a somewhat cooler climate might have come into being in the Jurassic, with further segregation and evolution during the Early Cretaceous of others better adapted to even more temperate conditions. In his excellent summary, based on paleontological evidence, Axelrod points out that by the Mid-Cretaceous this ancestral magnoliopsid flora had become widely deployed in the frost-free regions and had also given rise to groups characteristic of the northern and southern temperate zones. There is a suggestion that the Paleocene — the transition from the Cretaceous to the Tertiary — saw an even wider development of a characteristically pantropic flora. Interesting as was this deployment of the magnoliopsids during the Cretaceous — a large pantropic series, with considerably different forms in the north and south temperate regions — the forests of the time were by no means lacking in pinopsids. The latter were abundant in the cooltemperate regions and some have remained there as characteristic forest types to this day.

In the frost-free and warm-temperate parts of the southern hemisphere, broad-leafed pinopsids competed successfully with the magnoliopsids, as they still do over large areas; and forms apparently better suited to cool-temperate conditions, such as then existed on what is now the Antarctic continent, were also represented. In the northern hemisphere the lowland forests of the Early Cretaceous were primarily composed of pinopsids, with a few scattered magynoliopsids, whereas by the close of the Cretaceous the arborescent magnoliopsids had increased to about 60 per cent of the total known flora, the pinopsids had dropped to about 15 per cent and the remainder were mostly ferns, cycads and highly evolved herbaceous materials, such as sedges and grasses. There is considerable evidence, however, that, except in equatorial regions, the uplands of the Late Cretaceous still contained abundant pinopsids.

In mapping forest cover, foresters, as a justifiably practical expedient, usually designate the stands on the basis of the dominant species present today. This practice may, however, at times give an incorrect picture of the natural forest cover, especially where the stands are known to be in ecological imbalance because of disturbance. For example, it is today usually considered expedient to map large areas in the southeastern United States as being dominated by "coniferous forest."

However, as every forester in the region knows, the constant problem is to keep a stand of pine on the more productive sites from going over to a mixture with hardwoods which, to him, are "weed species." Only on exceptionally poor sites do the pines maintain their dominant position without assistance. This is clear evidence that the region basically is one of a mixed forest type. Today controlled burning is being ever more widely accepted as a legitimate silvicultural method for the control of hardwoods in the extensive southern pine forests. In a similar vein, evidence is only now accumulating that the once extensive stands of white pine in the New England states were the result of recurrent hurricanes which, periodically for a very long time, have swept the area with destructive force.

Under undisturbed conditions in New England hardwood forests dominate the stands, with white pine scattered, any pure stands of it being limited to rocky outcrops and patches of extremely sterile soil. On more fertile soils in New England white pine is merely the first stage in a normal ecological succession following extensive blowdown by hurricane winds, or after clearing or fire. Here, as in the southeastern states, mixed forests of magnoliopsids and pinopsids would be the natural climatic cover.

Thus, without going into details on a world-wide basis, we may here note that mixed pinopsid-magnoliopsid forests, exhibiting an infinity of variations from region to region, but always with much the same general physiognomy, are widespread under warm- to moderate-temperate conditions if the rainfall is adequate and the areas but little disturbed. In fact, when examined objectively, the so-called "coniferous" forests of cool-temperate regions usually have admixtures of hardwood species. As far as associations of species are concerned, these and similarly mixed forests, scattered across the temperate regions of both the northern and southern hemispheres, belong to what have been designated as various segments of the "Tertiary Forests"; but, as types of forest communities, they are more ancient and have their roots in the mixed pinopsid-magnoliopsid forests of the Cretaceous.

PRELUDE TO THE PRESENT (TERTIARY PERIOD)

By the end of the Cretaceous all the important families of magnoliopsids containing arborescent members were present and many of our modern genera were abundantly represented. The Tertiary marked the elaboration of species. In the Early Tertiary, the frost-free regions were somewhat more extensive than they had been in the latter part of the Cretaceous and offered ample opportunity for the basically warm-climate magnoliopsids to migrate. This probably explains the wide dispersal of so many tropical genera today, genera already in existence in the Tertiary.

It also applies to the genera which today are widespread or exhibit disjunct distributions in the north temperate regions (Holarctica). Similar disjunct distributions of a series of families and genera of both magnoliopsids and pinopsids in the more temperate regions of the southern hemisphere, which seem never to have existed elsewhere, however, offer an incompletely resolved problem to paleogeographers. By Mid-Tertiary time, the climatic pendulum was swinging back again. The frost-free zone was slowly shrinking, treeless areas began to appear in the polar regions and the subhumid zones exhibited considerable expansion, their inner portions becoming arid. This series of shifts culminated in the first of the Pleistocene glaciations. As is well understood, a succession of ice maxima occurred during the Pleistocene, with interglacial intervals when the climate appears to have been considerably more genial than it is today. It has been calculated that the last ice maximum, covering perhaps as much as 12 million square miles of the earth's surface, locked up so much water from the world's oceans that their level was lowered about 300 feet.

The fact that the melting of the ice still remaining in both hemispheres would raise the surface of the world's oceans perhaps as much as 150 feet above its present level, suggests that we may be scarcely more than half way through the present glacial regression. There is no point here in entering the lists with those currently arguing over the actual extent of disturbance of the climatic zones during the height of the Pleistocene glaciations. The shifts certainly were not so marked as during the Permian and the floras of the world appear to have recovered lost territory with considerable ease, except in certain areas (e.g., in Europe, where there was a considerable loss of those species and genera characteristic of the later Tertiary forests of that continent). Where there was ample room for migration, some species that appear to be closely allied to (or perhaps the same as) species in Late Tertiary strata moved southward ahead of the ice and since then have been migrating northward again. The more genetically plastic groups apparently returned after each ice maximum, with a considerable accumulation of variability leading to the production of new species.

The Forests of the Presen

We are now ready to take our theoretical representation of the climatic zones of a period of world-wide topographic and climatic disturbance and relate it to the vegetation types of the present. Instead of climatic zones, generalized vegetation types appear. Here, again, it should be emphasized that local conditions, such as extensive mountain ranges, plateaus, or the relation of land and ocean masses, modify the details of the pattern. The map however, shows that the generalities hold.

POLAR REGIONS

The north and south polar regions retain icecaps on the lands and considerable amounts of floating ice in the adjacent seas. Although vegetation occurs on summer snow-free areas (the precipitation is low and snow cover often very thin), it is stunted and presents the general aspect of *tundra.* Locally, treeless tundra may extend below 60° latitude.

TEMPERATE FORESTS

The transition from the tundra to temperate conditions in the northern hemisphere is characterized by the predominance of spruce-fir forest, called *taiga.* The southern tip of South America is too temperate to have a comparable forest type. On theoretical grounds the further transition southward from the taiga should witness a progressively greater incidence of broadleaf deciduous species; and in general this occurs.

In the North American Pacific region, however, the expected superhumid area, in conjunction with the mountains and generally higher elevations, has thrown the balance in favour of the retention of predominately coniferous forests, although many broadleaf genera and species are also found, especially in the lowlands. In the eastern part of North America, in Europe and in eastern Asia, the forests are primarily broadleaf deciduous, but coniferous elements of the north temperate forest also are abundant; this is the *Mixed Broadleaf-Coniferous Forest.* The mixed broadleaf-coniferous forest of the southern hemisphere has much the same physiognomy as that of the northern, but with its own series of magnoliopsid and pinopsid genera. The superhumid region appears in southern Chile, greatly reduced in extent by the Andean chains; some of its characteristic species are closely related to forms living in New Zealand and other antipodean regions, or to forms known from Tertiary deposits of Antarctica.

In neither hemisphere is the transition from the temperate to the frostfree region marked by any sudden change in forest physiognomy, although the transition type is sometimes called *warm temperate forest.* This forest type often contains broadleaf evergreen species in genera which, in the more familiar temperate forest, are characteristically deciduous; broad-leafed evergreen pinopsids, unknown in northern regions, sometimes are found associated with this forest type in the southern hemisphere. The occurrence of a fairly large number of relic genera in this transition zone may indicate that during the Early Tertiary this forest type was of much greater extent. This we should expect, for the generally warmer climate of the time would have produced a wider and even less clearly marked zone of transition between the temperate and frost-free regions.

ARID AND SEMIARID AREAS

The two central divisions of the drier areas of the world fall outside the scope of this work. The innermost, *Desert and Desert Scrub,* too dry to support anything more than a sparse vegetation, is encircled by short-grass *Steppe.* Surrounding the steppe, however, is a third segment which, vegetationally, is of interest here, for it contains arborescent material.

SUBHUMID TRANSITION AREAS

This outermost zone is transitional between grassland and forest. Near the western continental margins adjacent to or slightly within temperate regions, the rainfall occurs mostly in the winter months, the summer being hot and dry. Here a *Xeric Scrub and Woodland* is present, with trees along streams or in scattered, sparse colonies, usually accompanied by winter-green grasslands. This type of vegetation, best developed in the Mediterranean region (it is often called the "Mediterranean scrub" type) and in parts of California, also occurs in South America, southern Africa and Australia. As this same transitional band swings toward the continental interiors, it becomes subject to violent storms and irregular precipitation. The trees are at first limited to stream valleys, forming "gallery forests". With slightly more precipitation, groves of trees and thickets of shrubs begin to appear away from streams.

In the mid-continent of North America, the groves were once largely dominated by oaks, often bur oak *(Quercus macrocarpa)* and offered a welcome relief to those crossing the grasslands during the period of early settlement; stag-headed relics of some of these oak groves still stand near old homesteads, but most of them have been removed because of the danger of storm-thrown branches. In North America the shrub thickets often were dominated by thorny species of plum *(Prunus)* and hawthorn *(Crataegus).* The intervening grasslands contained relatively tall grasses and have been called *Prairie.* It is here suggested that the term "prairie" might perhaps be expanded to take in the entire vegetation complex of the region.

In its undisturbed phase the prairie is a transition grassland containing gallery forest and occasional groves of widely spaced trees, usually associated with more or less thorny shrubs. This is characteristic of the Asiatic, Eurasian and South American prairies as of those of midcontinent North America. Today there is little opportunity anywhere in the world to see typical prairie. Climatically suited to the raising of excellent grain crops, for the grains are grasses, the natural prairie areas of the world are almost entirely under intensive cultivation, so great is the world's need for food. The gallery forest and scattered groves have largely fallen because of acute needs for fuel and local building material, although there is evidence that many of the groves and some of the thinner gallery forests were eliminated by prairie fires set by indigenous nomadic hunters too early to have been seen by travelers in modern times. The "treeless prairie," in large part, may be the result of clearing and fire.

Extending into the frost-free areas, the "tropical" margin of this subhumid region becomes *Savanna* at about latitude 30°, perhaps more by definition than by any immediately visible change in vegetation. Here the precipitation strongly tends to be concentrated in the high-sun months, although there may be a second period of low-sun rainfall. The savanna is characteristically a tropical grassland, often disturbed by fire, with gallery forest along the streams and with scattered groves or occasional individual trees. Palms are not infrequent and the shrubby growths tend to be thorny. Typical *Thorn Scrub* and *Thorn Forest* usually appear adjacent to either savanna, or light tropical forest and often develop out of these types after disturbance.

FORESTS OF FROST-FREE REGIONS

Despite the larger land areas in the northern hemisphere, more than half of the world's remaining forests are to be found in frost-free regions. In the higher mountains of the tropics, especially where contiguous to the temperate regions, the vegetation is likely to contain genera often thought to be "characteristic" of the temperate forest, commingled with elements of more "tropical" groups. Here the rapid changes in elevation, coupled with the great number of micro-environments on rugged slopes, lead to a complex type of vegetation.

At middle elevations, the forests merge into those characteristic of the adjacent tropical lowlands. In a work of this compass it would be useless to attempt any description of these frost-free montane forests. Where the precipitation is abundant-as it often is, especially on the windward sides — a *Montane Rain-Forest Complex* develops, with the dominant species often succeeding each other in relatively narrow altitudinal belts; it is "complex" because each species is present but progressively less conspicuous in a succession of adjacent zones.

DISTURBED TROPICAL AREAS AND "JUNGLE."

There is a growing body of evidence that many areas of present thorn forest and related thorn scrub have developed out of both savanna and light tropical forest following disturbance, especially overgrazing. The great development of tropical savanna in many areas, however, has been accepted as a natural phenomenon by many plant geographers. As noted previously, natural savanna would be expected to occur in the climatic tension zone between tropical steppe and light forest. Nevertheless, on critical examination, it becomes evident that there is a proportional imbalance between savanna and light forest in various regions.

While this is no more true of parts of Africa than of other tropical areas, it is most evident there, with far more savanna and much less light forest than we should expect on theoretical grounds. The often sharp line between relatively heavy forest and savanna, with no apparent differences in soil or precipitation, indicates that the sudden change from one vegetational type to the other is unnatural. There now is a strong tendency among recent students to conclude that wide expanses of savanna, as now defined, have been caused by fire. If so, the question is whether this is a recent happening, or whether — in some areas — it may be a legacy of the antique past. Again we may turn to Africa for evidence, evidence that has been available for a long time, but has been largely dismissed as distorted folk lore.

The *Periplus* of Hanno, the Carthaginian, the Greek translation of which still survives, is an account of a remarkable voyage of exploration along the west coast of Africa about the year 520 B.C. Hanno relates that some time after passing the mouth of the Gambia River he stopped on a coastal island, where the night was made hideous by the sound of drums, gongs and some sort of wind instrument and terrifying by great fires.

Leaving the place in haste, Hanno sailed on, keeping Africa on his left. Day after day columns of smoke rose from the land and by night the whole country seemed to be in flames. Whether the explorers got no farther than Nigeria, or ventured to the Cameroons, as some maintain, is of little moment here. The main point for this discussion is that Hanno recorded having seen great fires inland in Africa for days on end as he sailed along its coast.

Although Hanno failed to comprehend their nature, these fires must have been due to the annual "burning of the bush" to beat back the forest and provide land for agriculture and grazing. Thus we have evidence that for at least two and a half millennia — and how much longer we shall never know — great forest tracts in Africa have probably been regularly despoiled by fires purposely set by man, much as they are today. The result has been a great expansion of savanna at the expense of the seasonally combustible light forest; and the same thing has unquestionably happened over large areas in the Americas, in India and in southeastern Asia.

Only within recent years have students working in tropical areas been able to divorce their thinking from ideas gleaned from previous study of the exceedingly overpopulated temperate regions and of the impoverished floras with which they are already familiar. Simplified concepts of forest ecology based on the almost diagrammatic floristic pattern of northeastern North America, or especially of northwestern Europe with its relatively few forest species, do not readily apply to the tropics, nor are the farms and pasture lands of the tropics laid out in neatly fenced plots. The number of species in a given area usually is far greater in tropical than in temperate regions; and agriculture in many parts of the tropics is fundamentally migratory, for several very definite reasons.

In the drier portions, catch crops planted during, or immediately following, the rains are the rule. Here clearing is primarily by fire and the fires, set in the periods between rains, often get out of hand, spread over great areas and enter the forest margin for some distance. This produces the often observed sharp line of demarcation between forest and savanna. It does not take a particularly dense human population to bring about these effects and man has been living in the equatorial regions of both hemispheres for much longer than is sometimes supposed.

In the lowland portions of the frost-free zone that have abundant precipitation-in rain forest and light forest-there is and for a very long time has been, a considerable population, especially near the watercourses, the "jungle roads." Here patches are cleared for the indigenous crops (and more recently for those that have been introduced) and the land is cultivated. Again, however, one familiar only with agriculture in temperate regions is prone to misinterpret the situation; the native farmer is not being indolent when he clears a plot of ground, cultivates it for a few years and then abandons it to weeds. It is the only course open to him.

The generally high soil temperatures do not lead to humus formation and soon the abundant rainfall so thoroughly leaches the soil of its mineral nutrients that it is necessary to abandon the plot after only a few crops. A new area is cleared, the rubbish burned where it lies (an effective method of scattering the mineral-containing ash) and a new crop planted. Unfortunately, modern plantation agriculture in the tropics far too often follows this same pattern. Under this system "fallowland" is a former cultivated area, perhaps only temporarily abandoned as no longer profitable because of the dearth of nutrient minerals in the soil and the high cost of fertilizers.

After abandonment, these cleared and briefly cultivated areas return to vegetation natural to the area, but not to the original forest type. Such tropical "weed forest" is the most rampantly growing sort of vegetation known. But remarkable as is this secondary forest, with its dense undergrowth and tangle of lianas, it is characteristic only of the disturbed areas of the region.

Descriptions of these "impenetrable jungles," usually written in awed and sometimes lurid phrases by early travelers, who necessarily had to stay close to the rivers, have become the classic texts from which we have taken our concepts of tropical vegetation for much too long a time. Breaking out of this river-margin "jungle" and complex secondary growth that follows cultivation, the traveller may walk, as the writer has, day after day in the almost trackless forest of the South American equatorial region, only rarely having to swing his machete-except, if he be a botanist, to collect some choice specimen. The undisturbed tropical forest may be complex and can contain a great abundance of species, but it is far different from the "jungle" of much of our literature.

8

Forests, Trees and Water in Arid Lands

Arid lands are among the world's most fragile ecosystems, made more so by periodic droughts and increasing overexploitation of meagre resources. Arid and semi-arid lands cover around one-third of the world's land area and are inhabited by about one billion people, a large proportion of whom are among the poorest in the world. Forests, trees and grasses are essential constituents of arid zone ecosystems and contribute to maintaining suitable conditions for agriculture, rangeland and human livelihoods. In providing goods (especially fuelwood and non-wood products) and environmental services to the rural poor and in contributing to the diversification of their household sources of income, forests and trees in arid zones boost poverty alleviation strategies and reduce food insecurity. Roughly 6 percent of the world's forest area (about 230 million hectares) is located in arid lands (FAO, 2001). Trees outside forests (scattered in the landscape, in arable lands, in grazing lands, in savannahs and steppes, in barren lands and in urban areas) have a vital role in arid lands, although it is difficult to assess their extent.

Availability of water – surface water, groundwater and air moisture – is usually the main factor limiting natural distribution of trees in arid lands, along with climate (rainfall, temperatures, wind) and soil quality. Each tree species is adapted to certain conditions and is located in its "niche". When optimal conditions are widely distributed, forests or shrubs may cover large areas. More often, limited by water scarcity, vegetation is concentrated where run-off can accumulate or where groundwater is accessible. This leads to the uneven distribution of trees and bushes, for example in striped bush (fragmented bush stands), riparian forests, the deepest channels of valleys (thalwegs) and oases, and isolated in the landscape. However, the natural distribution of vegetation has long been altered by human activities. Deforestation and degradation of tree and shrub formations (mainly through conversion to agricultural use) and overexploitation of forests and woodlands (through fuelwood collection and overgrazing) are among the major causes of soil degradation in arid areas. Furthermore, global warming is expected to result in rainfall decrease throughout most of the world's arid zones, which will lead to more severe water scarcity and increased desertification risks. Many methods for reversing deforestation, degradation and desertification rely on tree planting. However, before trees are planted it is essential to consider the water balance.

A TREND OF DECREASING FOREST COVER

DEFORESTATION

Conversion of forests for agricultural crops and pasture land is the main cause of the increasing deforestation in arid lands. In many places the prevailing shifting cultivation and crop/fallow systems are no longer possible and continuous cultivation of the same piece of land, often with no crop rotation, leads to exhaustion of soil fertility and the need for new lands. Degraded wooded lands which were formerly neglected are now actively deforested. Increased grazing pressure and unmanaged harvesting of fuelwood and other products also result in degradation and deforestation. The remaining forests and wooded lands are sometimes threatened by pest and disease outbreaks, although these are rare under extremely dry conditions. Forest fire is a constant threat in arid lands, although very large fires are rare compared with those occurring in other regions. Limited fuel accumulation due to high grazing pressure limits the extension of burned areas. However, fires cause considerable loss of forest, bush and tree cover, especially in the drier ecosystems, endangering ecological niches hosting relicts of forests of high biological diversity.

Desertification

The United Nations Conference on Environment and Development (UNCED, 1992) defined desertification as "land degradation in arid, semi-arid and dry sub-humid areas resulting from various factors, including climatic variations and human activities". Desertification is not an advance of existing deserts but is rather the effect of localized degradation of the land. It rapidly follows deforestation and soil exhaustion. Exposed to the sun, the wind and the rains, exhausted soils loose their organic matter and their structure while nutrients are leached away. Fine elements are blown into dust storms and sand grains become mobile and encroach on other lands through sheets and dunes. Overexploitation of forest, tree, bush, grazing land and soil resources has been increasing desertification.

Desertification is a worldwide problem directly affecting 250 million people, particularly in Africa where two-thirds of the continent is dry lands and deserts. However, more than 30 percent of the land in the United States is also affected by desertification. One quarter of Latin America and the Caribbean is deserts and dry lands. In Spain, one-fifth of the land is at risk of turning into deserts. In China, since the 1950s, sand drifts and degradation have taken a toll of nearly 700 000 ha of cultivated land, 2.35 million hectares of rangeland, 6.4 million hectares of forests, woodlands and shrub lands. Worldwide, some 70 percent of the 5.2 billion hectares of dry lands used for agriculture are degraded and threatened by desertification (FAO, 2007a).

Climate change effects on arid lands

Undisturbed forests are able to adapt to climatic and edaphic changes to a certain extent, but not over the long term: palaeo-botanical records indicate that past climate change destroyed prevailing vegetation types and promoted new types to replace the former ones. According to most predictive models, global warming will affect arid lands through temperature increase and rainfall decrease all over the world (with the exception of southwestern Latin America, where more frequent El Niño–Southern Oscillations are expected to lessen drought risks) (UCAR, 2005). The models predict increases of the frequency and/or intensity of droughts. Increased fire risks are also expected for the remaining forests and wooded lands. Increased temperatures lead to increased evaporation and more severe scarcity of water. All these trends lead to increased risks of desertification. In many places the vegetation already faces harsh conditions near the threshold of lethal temperatures. Any increase of these maximum temperatures will directly lead to irremediable vegetation loss.

The main consequences of climate change in arid lands will be a decrease of agriculture, rangeland and forest productivity, biodiversity, soil organic matter and fertility. This will worsen poverty and food insecurity. Populations will be forced to migrate. It is predicted that 135 million environmental refugees will leave their land by 2020 because of desertification, of which 60 million will be displaced in sub-Saharan Africa (FAO, 2007b). Already facing the lost productivity of natural rangelands, nomadic and transhumant herders may be forced to settle.

Concentration of herds around their new homes has already led to the disappearance of most of the vegetation cover around many settlements and around wells and other water sources that provide drinkable water for humans and animals year round. Policies to support settling of nomadic herders are weak in many countries. Another problem is related to the ageing of the tree population as a result of overgrazing of young seedlings, which impedes the natural regeneration of trees. Overmature trees progressively lose their resilience to climatic stress, so that a single climatic event can destroy a whole area of forest. For example, most of the *Acacia nilotica* forests of the Senegal River Valley died in the early 1970s after a severe drought.

Restoring vegetative cover of arid-zone lands can help mitigate climate change by increasing carbon uptake and storage, even if only a small amount of carbon per unit area will be sequestered. The area of arid lands to be restored is so huge that it constitutes a good potential sink for carbon. The economics of related schemes should, however, be carefully considered and documented.

REVERSING THE DEGRADATION TREND

REMOVING THE CAUSES

To start with, the human-induced causes of desertification should be tackled. Poor people are obliged to exploit whatever resource they may have access to for their survival. Overexploitation should be avoided through assistance to meet their basic needs with income generation opportunities. Poverty mitigation measures can include planting trees (for their products and services) in major afforestation schemes, woodlots, linear plantations, windbreaks and hedgerows and as isolated trees in agricultural and other landscapes.

Natural regeneration through land protection

The most evident way to restore vegetative cover is to protect it from the causes of degradation: mostly exploitation (harvesting and grazing) and fires. Vegetation can spread naturally, even on bare lands, but the process is often slow. Protection is not always easy as it has to be maintained carefully over a long period. Planting trees, bushes and grass will speed up the process. Then, the restored lands need to be sustainably managed.

The Abéché protected area in Chad is a noteworthy example. In 1961, 305 barren hectares with a few *Acacia* trees (*A. raddiana*, *A. senegal* and *A. mellifera*) were fenced off with barbed wire and carefully watched over to protect the watershed. Within ten years, without any planting, total land cover had been obtained. After 45 years of almost continuous protection, the protected area is now clearly differentiated from its surroundings in satellite images.

Afforestation, sand dune fixation and green belts

Afforestation through tree plantation can be a good tool for environmental restoration. During the second half of the twentieth century many forest plantations were established in arid lands all over the world, mostly for protection or for fuelwood production, and the pace of plantation programmes has been speeding up (FAO, 2006a,b). Plantation programmes have used many species (often exotics) and techniques, from low investment (rainfed) to high investment (rainfed with land shaping or irrigated from the water table, deep aquifers or wastewater). The varied successes and failures of such plantations now constitute good sources of information for future activities. Many countries around the world (*e.g.* Chile, China, Denmark, France, the Islamic Republic of Iran, Mauritania, the Niger, Senegal and Viet Nam) have developed tree plantation techniques for fixation of shifting sands. In arid zones, both local and large national or international schemes apply such techniques to protect productive lands, infrastructures and settlements. Many of the plantations also produce wood and non-wood products.

Many arid zone towns and cities have planted local green belts to protect their population and infrastructures against dust storms and encroaching sands and to influence the microclimate. Arable lands, irrigation schemes, railways, roads, canals and coastal dunes are also being protected through dedicated schemes.

Larger-scale afforestation schemes for land reclamation have a long history; they were implemented in France and Germany in the eighteenth and nineteenth centuries and in the United States after the 1935 Dust Bowl. In Algeria, FAO and the World Food Programme (WFP) started the tree planting programme "Chantiers populaires de reboisement" in 1966. In 1971, Algeria initiated the "Barrage vert", a planted 20 km–wide belt on the fringe of the Sahara desert intended to stretch 1 500 km from the western to the eastern borders of the country, to comprise 3 million hectares. By 2003 only 100 000 ha had been planted, however, mainly with *Pinus halepensis* (Belaaz, 2003). Following this national initiative, North African countries (Morocco, Algeria, Tunisia and the Libyan Arab Jamahiriya) started a regional programme, the Arab Magreb Union (UMA) green belt for the north of the Sahara, but since the 1990s there has been little evidence of its activities. In 1978, China initiated the "Great Green Wall" project which afforested 9 million hectares in its first ten years. In the current phase of the project, now called "the New Great Wall", an additional 5 million hectares are to be planted by 2010 (Ratliff, 2003). Dust storms still trouble Beijing as yet, but airborne dust is carried such distances that the effects of such greening efforts may require several decades to become evident.

The African Union launched a "Green Wall for the Sahara" project in Abuja, Nigeria in December 2006 to contribute to halting and reversing desertification at the southern and northern fringes of the Sahara. The project will work hand in hand with all concerned countries and other organizations and programmes such as the New Partnership for Africa's Development (NEPAD), the Global Environment Facility (GEF) Operational Programme on Sustainable Land Management (OP 15), the United Nations Convention to Combat Desertification (UNCCD) and the TerrAfrica Initiative. Rather than merely establishing a few lines of trees, the project will address sustainable and integrated resource management and restoration activities (through tree planting, rangeland restoration and agriculture, implemented only where feasible and sustainable) in a land belt as wide as possible. It will be a task for several generations.

The results obtained from green belt experiences have varied greatly depending on the scale of the afforestation schemes, the quality of the methods applied, their adaptation to local conditions and the quality of the plantation management. An in-depth study of the climate, soil, water, land use and socio-economic conditions is always required. Local water availability and water demand must always be considered (see below). Green belt initiatives must also take into account previous land uses and ownership and the causes for deforestation and desertification, including peoples' needs for forest products, pasture and croplands, offering alternative solutions to cover these needs. Local people should be involved all through the process, from conception to the management of the new resources. Large monospecific tree stands should be avoided and a patchwork of different types of plant cover (including agricultural crops and grazing) preferred whenever possible. Local species should be preferred; several projects have clearly shown the problems related to exotic species, which may become invasive in their new environment.

EFFECTS OF OZONE ON FOREST TREES

FOREST RESOURCES

The forested areas within the SAMI region cover more than 50% of the land area and are very diverse in nature. Most of the tree species in the region are deciduous hardwoods with various species of oak, hickories, maples and yellow-poplar (*Liriodendron tulipifera*) as the predominant species. Many other species of hardwoods do commonly occur in the region. Although not as prevalent as hardwoods (< 30% of the total wood volume) several different coniferous species can commonly be found in the SAMI region. These include eastern hemlock, Fraser fir, red spruce, pitch pine, Virginia pine and loblolly pine. A more detailed description of the forested resources in the region is presented in *The South's Fourth Forest: Alternatives for the Future* (USDA Forest Service 1988).

Natural Stressors

Besides air pollutants, many other stresses affect forests in the SAMI region. These can be found in any forest in the world and include abiotic agents such as moisture and nutrient deficiencies (Dougherty 1995) and biotic stresses such as insects and diseases (Hepting 1971, Skelly *et al.* 1987). Commonly found insects include the balsam wooly adelgid (*Adelges piceae*), gypsy moth (*Pothetria dispar*), several bark beetles (primarily *Ips spp.*) inclusive of the southern pine beetle (*Dendroctonus frontalis*), and diseases such as sycamore anthracnose (*Gnomonia veneta*), Armillaria root rot (*Armillaria spp.*), numerous cankers, fusiform rust (*Cronartium quercuum f. sp. fusiforme*), littleleaf syndrome (*Phytophthora cinnamomi*), and annosum root rot (*Heterobasidion annosum*). Although primarily endemic in nature, insect attacks and diseases can reach epidemic proportions in the SAMI region as evidenced by the balsam wooly adelgid, gypsy moth, and the fungi which cause chestnut blight [*Cryphonectria (Endothia) parasitica*], dogwood anthracnose (*Discula destructiva*) and butternut canker (*Sirococcus clavigignenti-juglandacearum*). Wildlife activities also cause serious losses in the forest with particular concern for damage to regeneration by the whitetail deer. In addition, numerous abiotic stressors inclusive of local and/or regional scale major and minor periods of drought, rime ice and more heavy accumulations of ice leading to top breakage, freezing temperatures leading to winter injury, flooding, and of course fire are all present within the forests of the SAMI region. Moisture and nutrient stresses can be limiting factors in both the productivity and species composition of the various forest types in the SAMI region.

The interrelationships of ambient ozone exposure with all of these natural stressors remains virtually unknown. Introduced insects and pathogens usually need only a susceptible host specie(s) and favorable growing conditions for rapid speed and reproduction into epizootic/epidemic proportions. More is known about other abiotic stressors such as with drought and/or other pollutant interactions as reviewed later in this document. These naturally occurring and introduced stressors as may be found throughout the SAMI region are described in more detail by Skelly et al. (1987), Meadows and Hodges (1995) and Tainter and Baker (1996).

Objectives

The synthesis document will use information obtained from the annotated bibliography (already submitted to SAMI) to assess the scientific evidence concerning ozone effects on forest trees (primarily Class I areas) in the SAMI region. The primary emphasis of the document includes seven items: 1) evaluation of ozone effects on visible injury, growth and productivity, and physiological function; 2) review and classify historic trends, status and contribution of ozone to these trends in the SAMI region; 3) evaluate ozone exposure-response relationships derived from both greenhouse and field studies; 4) critique existing ozone exposure data sets applicably to Class I areas and their potential for use in assessing status of Class I areas; 5) evaluate other abiotic and biotic factors that may influence forest response to ozone and assess the potential influence of ozone compared with other environmental and plant factors; 6) review alternative interpretations of existing data; and 7) identify information gaps that limit assessment of current and future status of forest resources in the SAMI region. Recommendations from this assessment will be used by SAMI to address three policy questions:

- What is the current status of the resources relative to ozone effects on vegetation?
- What is the relationship between ozone exposures and vegetation response?
- What changes in resource status are projected to occur from changes in exposures due to implementation of the 1990 Clean Air Act Amendments or other emission management options being considered by SAMI's Policy Committee?

VISIBLE FOLIAR SYMPTOMS

This introductory section, as adapted in large measure from Skelly *et al.* (1987), will present descriptive information on the visible foliar injuries as induced by exposures to ambient ozone under natural field conditions within the SAMI region. It is recognized that pre-visual physiological changes are induced before the manifestation of such visible symptoms. Growth and/or productivity changes may also precede or follow visible symptoms. Please refer to the respective sections that review these forms of pre-visual and productivity changes.

Visible symptoms resulting from ozone exposure are generally recognized as resulting from either acute or chronic exposures as manifested on the foliage of sensitive plants. Injury from acute [unusually high concentrations (generally >0.10 ppm in the SAMI region), short-term (generally <24 hr)] exposures normally involves the death of cells and develops within a few hours or days following exposure. Injury is expressed as tissue bleaching, leading to bifacial necrosis (fleck), and may follow exposure to unusually high ozone concentrations. Acute ozone exposures rarely occur in nature.

Foliar injury resulting from chronic exposures [lower concentrations (generally < 0.100 ppm in the SAMI region), longer-term (generally days, weeks, or months)] may be manifested as chlorosis, pigmentation (stippling), general leaf reddening and/or premature or enhanced senescence. Both types of symptoms (resulting from acute or chronic exposures), particularly those from chronic exposures, may be confused with symptoms of other conditions, such as normal senescence, nutritional disorders, other environmental stresses, biotic pathogens, or insect infestation (Skelly *et al.* 1987).

Description of Symptoms

Broadleaf Species

Under conditions of ambient ozone exposures several specific symptoms appear on broadleaf species in the field, the most common of which is stipple, defined as discrete areas of pigmentation restricted to the adaxial leaf surface. The upper leaf surface of sensitive plants may have a tan, red, brown, purple, or black stippling (as if pepper had adhered to the upper leaf surface) that may appear uniformly over the leaf surface, or restricted to certain areas of the leaf, or discrete dot-like lesions.

The lower leaf surface usually remains uninjured, with only the palisade cells underlying the upper epidermis affected. Veins and veinlets are usually not involved, and small veinlets often bound the injured areas, producing angular sections of affected tissue. Sometimes stippling is best observed by holding the leaf toward the sun; symptoms are often more intense on leaves exposed to direct light. Recent evidence suggests that shaded leaves may be more sensitive earlier in the growing season and exhibit the earliest symptoms (Fredricksen *et al.* 1995).

Stippling has been described as the classic symptom of ozone injury on broadleaf trees. The coloration of stippling is usually characteristic for a species, but can vary with environmental or physiological conditions. The youngest fully expanded leaves are normally the most sensitive, although all but the youngest leaves may be affected. On young leaves, symptoms tend to develop at the tips, on older leaves, toward the base. The entire surface of older leaves may exhibit symptoms when exposed to ozone periodically during the growing season. Although stippling is usually interveinal, following repeated and continually higher ozone exposures, extensive injury may also occur to the leaf veins themselves resulting in generalized chlorosis. In some species, especially sycamore, stippling tends to appear adjacent to the larger veins. Chlorosis, or loss of chlorophyll, may occur as a generalized condition similar to senescence, in discrete patches called mottle, or in patterns similar to stippling. Chlorosis is often more prevalent on the upper leaf surface of species that have palisade cells. On plants with pinnately compound leaves, such as ash, hickory, and tree-of-heaven, only some leaflets at certain positions along the petiole may be affected. Because stippling is a photosensitive response, overlapping leaflets or leaves may create a sharp line of demarcation between injured and uninjured (shaded) tissue.

On broadleaf species that are very sensitive to ozone, injury is usually expressed as a fleck characterized by small, discrete areas of dead cells-usually palisade parenchyma and sometimes associated epidermal tissues-leading to the formation of irregular lesions that although first evident on the adaxial leaf surface injury quickly become evident on both leaf surfaces. The lesions are often bleached (unpigmented) but can be colored as in stippling, and the affected areas may be slightly sunken. Bifacial necrosis results when the tissues connecting the upper and lower leaf surfaces are killed. The tissue coloration ranges from white to red-orange to black and is often characteristics of a specific tree species. Small veins are usually included in the necrotic tissue, but larger veins often remain alive. The area of dead cells collapses, and the upper and lower leaf surfaces adhere together to form a papery lesion.

In bifacial necrosis, the leaf margins are sometimes the most severely injured. Prolonged exposures to even low ozone concentrations may cause coalescence of chlorotic tissues, light stippling, and/or production of a bronzed appearance, especially on ozone-sensitive tree species.

Season-long exposures to ambient ozone may further result in premature or enhanced senescence of the older leaves on sensitive species. This may be a classic symptom of cumulative ozone exposures for many species, but remains elusive of description due to lack of filtered air controls for purposes of phenological comparisons. For sensitive species such as black cherry, notable stippling and general leaf reddening of the older leaves has been observed, with defoliation occurring several weeks before more tolerant individuals on the same site (Skelly, personal communication). For other sensitive individuals within species such as yellow-poplar (*Liriodendron tulipifera*, L.), white ash (*Fraxinus americana,* L.), or sassafras (*Sassafras albidum* L.) premature senescence is characterized more simply by the yellowing and dropping of the older leaves. When defoliation is severe, leaves with ozone-induced stipple may commonly be found on the forest floor below sensitive individuals several weeks before full autumn coloration.

Conifer Species

Foliar injuries on conifers as may be induced by ambient ozone exposures are far more difficult to identify in comparison to the more distinctive symptoms induced on broadleaf species. Although ozone induced symptoms were once thought to be easily diagnosed, recent controversies (Skelly, personal communication) about the causes of several needle anomalies have added to the confusion of proper diagnosis, especially under ambient forest conditions and following ambient ozone exposures. So many other causes may be involved in inducing the symptoms described below that determination of cause/effect relationships during field surveys may prove difficult.

Usually, on conifer species the two most common needle symptoms are chlorotic mottle and tipburn (Skelly *et al.* 1987). Overall, mottle of young and older needles may be induced by low ozone exposures, while tipburn of young needles may be produced by higher, early season exposures. There is, however, considerable difference in symptom expression depending on species, within-species variation, timing of exposures, and environmental conditions. Chronic symptoms usually only develop during the late summer or early fall on the current year needles with chlorotic mottle and/or spotting developing as small patches of yellow tissue interspersed with green areas (mottle), or as discrete yellow spots surrounded by apparently healthy tissue. Chronic symptoms in two, three and older-year needles simply intensify with coalescing of symptomatic areas leading to more general needle chlorosis and early senescence. Loss of second and third-year needles on eastern white pine (*Pinus strobus*, L.) and loblolly pine (*Pinus taeda*, L.) following advancing chlorotic mottle, spotting, and general chlorosis has been reported (Skelly *et al.* 1987, Chappelka and Chevone 1992).

Symptoms on most conifer species occurring as the result of acute exposures are rare and usually occur following episodic events to ambient ozone early in the growing season. If an ozone episode were to occur at this sensitive time of needle growth, that portion exposed to the higher ozone concentrations may subsequently develop necrotic needle tips resulting in tip dieback or necrotic banding. Tip necrosis is characterized with a pink to reddish coloration that fades brown with age. Towards the end of the growing season, these necrotic tips may break off, making affected needles appear much shorter. The tipburn symptom, especially in eastern white pine, usually affects all needles in a fascicle equally following an exposure to episodic ozone in the spring portion of the growing season. Severely affected individuals are evident within populations due to their off-color for much of the growing season.

On eastern white pine, a late season phenomenon of banding in the mid-portions of individual needles within the current year fascicles has also been attributed to episodic ozone exposures immediately before symptom expressions. Although shown to be induced by high ozone exposures under controlled conditions, additional controversy as to the role of fungi in causing this symptom leads to a less clear decision of determining this symptom as solely being due to ambient ozone exposures under forest conditions (Skelly, personal communication). Young, rapidly growing needles that are directly exposed to sunlight are the most sensitive, but older needles can exhibit mottle and premature senescence from prolonged, chronic exposures. Similar to broadleaf species, symptoms on plants within a single conifer species can vary considerably.

Bioindicator Species

Some particularly sensitive broadleaf tree species include black cherry, white ash, yellow poplar, and sassafras have become recognized as very useful bioindicators of ozone exposure because of a distinctive upper surface stipple, leaf reddening, and a large portion of the population appearing as sensitive under varied conditions of exposure and environments (Chappelka *et al.* 1992). Clonal lines of eastern white pine have been suggested for use as bioindicators of ambient ozone exposures, but as described above the general use of this species under field conditions may pose difficulties in determining cause/effect relationships. Other forest species that may serve as bioindicators of ozone pollution are blackberry, milkweed, dogbane, big-leaf aster and poison-ivy. Blackberry and poison-ivy exhibit a dark purple-red stippling of the upper leaf surface that often coalesces over most of the leaf surface. The symptom on milkweed is similar to that on blackberry, except that the coloration is purple-black. Milkweed has been reported as a common bioindicator of ozone (Duchelle and Skelly 1981).

IMPROVING THE WATER BALANCE

Natural forests and tree plantations improve the water cycle in diminishing run-off and improving the replenishment of the water table. Tree planting has often been proposed as a way to increase rainfall. It has been estimated that 60 percent of rainfall over the moist evergreen Amazon forest comes from the forest itself through evapotranspiration (TheAmazon.org, 2007). However, planting trees will produce tangible results in increasing rainfall on neighbouring areas only when very large areas are converted to forest (Avissar and Otte, 2007).

However, trees also consume water. The more the aerial system of trees is developed, the more water they transpire. The desirability of tree planting in arid lands is debated because trees may consume more water than they provide to the water cycle. Some countries, such as South Africa, have imposed a tax on the water consumed by forests. In certain circumstances where trees consume all the rainwater, it may be judged better to harvest this water through a bare watershed, store it in a reservoir and use it to irrigate high-value agricultural crops. For example, in Yatir, Israel, where average precipitation is only 270 mm per year, more than 3 000 ha of rainfed *Pinus halepensis* were planted in the early 1960s under a large-scale afforestation project. Although the forest provides carbon sequestration benefits and contributes to the livelihoods of nearby communities (particularly through fuelwood and non-wood forest products such as resins, fodder and medicinal and aromatic plants), it uses all the precipitation water. Furthermore, the forest has altered the biodiversity of the region, as new predation dynamics threaten endemic species. Rueff and Schwartz (2007) reported that the water that the watershed would have provided if it had not been afforested would have alleviated poverty better if it had been used for agriculture. They suggested that afforestation on a smaller scale, such as on farmers' plots, may yield similar benefits with fewer drawbacks, as combining tree planting and agriculture is less disturbing to the environment, improves agricultural yields, conserves water and soils and provides fuelwood for farmers.

Local populations have implemented different methods of water harvesting to benefit their crops and their trees. One technique was adapted from the natural example of the striped bush at the transition between continuous bush stands and grass steppe (Malagnoux, 2008). Where rainfall does not provide enough water to maintain a continuous vegetative cover, fragmented vegetative cover is separated by land strips of varying width. Run-off from the bare land strips provides the vegetation with the water it needs; the strips thus constitute small watersheds. Agronomists have improved on these traditional techniques, and foresters have also adapted them to the size and needs of their trees. Mechanized technologies have been developed to increase the scale of land restoration dramatically through quicker and cheaper land processing, while deepening the strips to increase water-holding capacity.

In tree planting for desertification control, the present and future water balance of the stand should be systematically estimated for each phase of its evolution. Appropriate silvicultural measures should be promoted to maintain the yearly water consumption below the yearly water inflow – including species choice, surface to be planted, planting density, thinning, pruning, coppicing, pollarding and also, when necessary, conversion to more sustainable vegetative cover, for instance from a dense stand to a parkland or grassland. Every desertification control programme or greening activity should be considered at the landscape level. Trees should be planted only when really needed and where possible.

In addition to rain, other water sources such as recycled water and deep aquifers have to be considered. Many arid lands and deserts have deep aquifer resources that could be tapped. While some restoration activities could rely for a short period on fossil aquifers, these activities will be sustainable only when water recharge exceeds or equals water withdrawal. With urbanization accelerating in arid lands, urban forestry and other urban greening programmes using vegetation that consumes less water than trees (*e.g.* bushes and grasses) are of increasing importance. More recycled water is being used in such programmes, including sewage water in some countries, and this practice will progressively grow in the future.

TROPICAL FORESTS

Primary tropical forest is not easy to find, although much has been so labeled by certain writers. In his excellent treatise on the world's tropical rain forests, Richards adds a most revealing footnote. In discussing what, in the original text, he characterizes as "Primary Mixed Forest" in Nigeria, he remarks that further work in the area "makes it seem likely that this forest has suffered disturbance in the past and is old secondary rather than truly primary." He adds: "The same is probably true for nearly all so-called virgin or primary forest in Nigeria and perhaps the whole of West Africa." Undisturbed primary forest is usually thought to be of considerable extent in the American tropics; but it is by no means so widespread as is generally supposed.

THE *TROPICAL RAIN*

Forest is perhaps our least understood vegetation type. Many forest areas in the tropics classed as "rain forest" actually are *Light Tropical Forest.* A further complication is the presence of large forest tracts on mountain slopes in the tropics which, because of the terrain and wind pattern, are well watered. These are *Tropical Montane Rain Forests;* on the middle slopes where the precipitation is usually highest, they have a composition considerably different from that of the rain forests of the lowlands. Where they are adjacent, both lowland rain forest and light forest pass into the montane rain-forest type.

However, if we take a rather narrow zone across the equator, including those areas in which rain forest would be expected to develop, we discover that it is advisable to particularize our information. It is immediately obvious that there is only a narrow strip along the equator in which rain can be expected to fall at almost all times of the year. Here seasonal maxima may occur, but only exceptionally does any long period pass without rain. This relatively narrow equatorial band is the locale of the true rain forest, although local conditions in some places produce equivalent conditions that enlarge the occurrence of this forest type. Furthermore, some of the areas commonly mapped as rain forest actually contain considerable tracts of light forest.

Wet and dry seasons begin to be evident at only a few degrees of latitude north and south of the equator. The total annual rainfall may be very great, but, on the whole, true rain forest does not exist where the dry seasons last longer than a few weeks. However, in such areas along the main river valleys and even along the smaller streams, the residual soil moisture is such that what might reasonably still be classed as rain forest does occur.

This readily explains why observers traveling only along streams have in the past often expanded the supposed general distribution of this type of forest far beyond its actual limits. Glimpses of forest were seen in the distance and it was supposed that it was of the same general composition as that near at hand. Actually this riparian "rain forest," intercalated with light forest, is comparable to the fingers of gallery forest that follow the streams in savanna regions. Light forest is found in those regions near the equator where there are two yearly maxima of rain, with one of the dry seasons of about three months in length. With a shortening of the longer dry season, this forest type approaches the rain forest in character.

However, on well-drained to overdrained terrain, the amount of moisture retained-the "effective precipitation" — may be low enough to produce characteristic light forest relatively close to the equator (and, in fact, in some places actually athwart it). Where the rainfall pattern is such that there is only one rainy period and a correspondingly long dry period, light forest persists until the conditions are reached where natural savanna (or in some instances thorn forest and scrub) occurs. The light forest is obviously not a single forest "type." It may be classified on a local basis, but in broad pattern it varies from a "wet" phase to a "dry" phase through myriad nuances of vegetational association. In discussing both the rain forest and what actually is the "wetter" phase of light forest, Richards indicates something of its nature. There are exceedingly complex ("mixed") types, with no species really dominant in the stands and there are also extensive forests with single dominant species.

He likens the "mixed" tropical forest to the mixed mesophytic forest of eastern North America, which develops only in areas with abundantly varied micro-environments. I suggest here that the "mixed" forest of the tropics may also be a type that develops primarily in areas with a series of varied micro-environments, as yet essentially unstudied and therefore not understood. The forests near major streams, where they are subject to periodic, or occasional, inundation — areas often many miles in width — are, of course, "disturbed forests" and therefore likely to be complex, with a characteristic mixed community and a considerable amount of "jungle" undergrowth. In the light forest, unless disturbed in some manner, the canopy is usually sufficiently close to prevent herbaceous and shrubby forms from receiving light enough to form more than a hint of jungle-like undergrowth.

On the whole, agriculture in the tropics is not successful in those areas characterized by typical rain forest, nor is life easy there. Hence, in his constant search for habitable places, man long ago entered the light forest, cleared the land and set about his routine chores of cultivating it. Great areas in India, once covered with "monsoon forest" (a typical form of light forest produced by the strong periodicity of rainfall), are now given over to intensive agriculture. The more primitive pastoral agriculture of central Africa has also eliminated large stretches of light forest, but here savanna has resulted, so that it is today most difficult to distinguish between natural savanna and that which has been artificially produced by fire and overgrazing.

Great as are the forest resources of the tropics, we have as yet made far too little intelligent use of them. At present, the world's primary demand is for the coniferous softwoods. Numerous especially valuable timbers are today being cut from tropical forests, but the great majority of the species are ignored. Remote from the daily experience of most of us, the spoilage of the remaining areas of tropical rain forest and light forest through improper practices is not generally comprehended.

Under our present economy many of the species of these forests are considered "worthless" and their normal habitats are sometimes almost wantonly destroyed. These "worthless" forms constitute at least a vast reservoir of raw cellulose; their other uses may be legion, should we study their characteristics in greater detail. Here is a great natural resource rapidly being wasted away through ignorance.

Variability and Kind

The origin of forest types, those times in the history of the world when the land surfaces were elevated, when mountain building was at a maximum and when climatic disturbances were widespread marked important periods in the development of new vegetation types. In view of the changes that land surfaces and climates constantly undergo, a too rigid genetic system is a distinct disadvantage to any group of organisms. With genetic plasticity — variability - there is a greater chance of survival during any disorganization of habitat. Variability, indeed, is the basis of survival. From the standpoint of plant systematics the species is usually considered to be the working unit. Furthermore, it is traditionally supposed to exhibit but little variation. This concept, however, is largely based on superficial morphological grounds.

As we begin to study species in greater detail, examining thousands of individuals instead of only a few in each species, it becomes evident that even this so-called morphological constancy of the species has little basis in fact; we have not been scrutinizing our materials with sufficient precision. We also are learning that, with this morphological variability, there is an equally large amount of physiological variability, inherent in the genetics of the group; and, in the end, this inherited physiological variability is far more important than are differences in morphology where factors concerned with survival are involved.

Foresters have long known that, in reforestation or afforestation, it is best to use seed from sources not too far removed from the site of the new plantings. In western North America Douglas fir (*Pseudotsuga taxifolia*) grows from about latitude 55° in British Columbia to about latitude 170 in Mexico, from sea level in the Pacific Northwest to elevations of over 10,000 feet in the Rocky Mountains and from humid to subhumid habitats. Attempts have been made to set up a series of "segregate species" from within the known morphological variability of this wide-ranging tree, but from the point of practical botanical taxonomy only the variety *glauca* is now recognized apart from the "typical" form. However, it has been found best not to move materials much more than a degree of latitude, or a climatically comparable distance in elevation, if the greatest survival and maximum growth of the planting is desired, though practical considerations usually make this impossible to achieve.

As a species defined by taxonomists, Douglas fir is widespread, but each of its local biotypes has a genetic norm that fits it to a particular set of ecological conditions. Even here we can discover considerable variability, related to the precise conditions of the local micro-environments. There also is a growing body of data that day length at certain periods of the year — which, of course, is correlated with latitude — plays a very important part in the effective competition and survival of these minor biotypes.

These and similar examples might be greatly expanded. The white pine (*Pinus strobus*) is found from eastern Canada through the Appalachian Mountains and also in the highlands of southern Mexico. When brought together at about latitude 40° N and if protected by appropriate handling under glass, the material from Mexico grows considerably faster than does that from the central Appalachians, as in turn this grows faster than do the Canadian plants. When planted in the open at this same latitude, the more nearly local material still grows faster than does the Canadian material, but the Mexican plants die with the first hard freeze. Data of the same sort are being accumulated for hardwoods that have wide distributions and also apply to adaptability to variant soils as well as to differences in climate-in brief, to the host of factors that influence the growth and persistence of plants.

The source of this genetic variability within species and the mechanics of its operation within species populations, which often enables the sum of their biotypes to have considerable geographical ranges and ecological tolerances, is a subject too extensive to be developed here. Nevertheless, what we can ascertain by relatively precise experimental methods with regard to what is happening among the species around us today, is indicative of what has been going on through the long span of time during which forests have covered the land.

Through such studies we come to an understanding of how it was that, in the past, groups of organisms could develop biotypic variations capable of venturing into new habitats and new climatic regions. Thus we glean insight into the importance of the uplands and mountain ranges of the past. It was on the uplands and cooler mountain slopes of the frost-free regions that the ancestors of our modern forest trees evolved. On these same slopes, with their multitude of different micro-environments, variant forms better preadapted to conditions in temperate regions came into being, were further elaborated, could persist for long periods of time and so be available to spread poleward during those periods when the great cyclic changes in the topography of the continental masses and their climates afforded an opportunity.

Likewise, this same inherent tendency toward variation, characteristic of all life, gave rise to forms still better fitted to cooler or even to boreal climates. With a swing of the climatic pendulum in the other direction, the ancestral biotypes would necessarily have had to retreat, or, if unable to do so for some reason, become extinct, leaving behind the newly evolved forms. These derived forms are the backbone of the vegetation of our present temperate and cool-temperate forests. With further evolution, coupled with genetic isolation, these biotypic segregates could — and have — become the "boreal" and "extreme austral" species of the taxonomist.

There is a tendency of some people to think of the tropical lowlands, because of their generally equable conditions, as having had a persistent and stable series of forest communities over long periods of time. This would seem to be a valid assumption on theoretical climatological grounds and also because considerable segments of the equatorial lowlands have been in a reasonably stable geographical position, at least since the Permian. However, there is evidence that the present forest types of the lowlands have been derived in large part from more upland types, which migrated into the lowlands during the Middle and Late Tertiary. As the writer has pointed out elsewhere, the forests of the central Amazonian basin appear to be primarily composed of elements derived from the surrounding uplands, perhaps at a time no earlier than the Late Pliocene or Early Pleistocene.

Obviously, based as it is on the consideration of a single area, this generalization needs considerable study before it can be taken as being more widely applicable. On the whole, however, the forests of tropical lowlands do not contain what, according to generally accepted botanical criteria, are considered to be the more primitive living species of genera that have fairly broad altitudinal distributions. These are usually found on the piedmont and lower mountain slopes.

Proper utilization of our forest resources demands that trees be cut down. What comes back to take their place-or what is brought in to replace a forest area that has been despoiled by improper cutting practices — is not a thing to be decided in a casual manner. Even with selective thinning, the micro-environments have been changed and with clear-cutting a totally new set of factors is introduced. Some few forest types regenerate satisfactorily; others are replaced — or might better be replaced — with different sets of materials. Whether the replacement is to be with the same biotypes, with somewhat different biotypes of the original species, or with entirely different species, depends on a considerable array of factors.

These factors are within the province of scientific silviculture, but they cannot be ascertained without a great deal of painstaking research.Biotypic variation is the fundamental mechanism leading to the evolution of the manifold kinds of organisms with which this world abounds and an understanding of it is indispensable in any consideration of the proper utilization and regeneration of our forests, our most important replaceable natural resource.

9

Genetics of Plant Breeding

Plant breeding is the art and science of changing the genetics of plants for the benefit of mankind. Plant breeding can be accomplished through many different techniques ranging from simply selecting plants with desirable characteristics for propagation, to more complex molecular techniques Plant breeding has been practiced for thousands of years, since near the beginning of human civilization. It is now practiced worldwide by individuals such as gardeners and farmers, or by professional plant breeders employed by organizations such as government institutions, universities, crop-specific industry associations or research centers. International development agencies believe that breeding new crops is important for ensuring food security by developing new varieties that are higher-yielding, resistant to pests and diseases, drought-resistant or regionally adapted to different environments and growing conditions.

DOMESTICATION

Plant breeding in certain situations may lead to the domestication of wild plants. Domestication of plants is an artificial selection process conducted by humans to produce plants that have more desirable traits than wild plants, and which renders them dependent on artificial (usually enhanced) environments for their continued existence. The practice is estimated to date back 9,000-11,000 years. Many crops in present day cultivation are the result of domestication in ancient times, about 5,000 years ago in the Old World and 3,000 years ago in the New World. In the Neolithic period, domestication took a minimum of 1,000 years and a maximum of 7,000 years. Today, all of our principal food crops come from domesticated varieties. Almost all the domesticated plants used today for food and agriculture were domesticated in the centers of origin. In these centers there is still a great diversity of closely related wild plants, so-called crop wild relatives, that can also be used for improving modern cultivars by plant breeding.

A plant whose origin or selection is due primarily to intentional human activity is called a cultigen, and a cultivated crop species that has evolved from wild populations due to selective pressures from traditional farmers is called a landrace. Landraces, which can be the result of natural forces or domestication, are plants (or animals) that are ideally suited to a particular region or environment. An example are the landraces of rice, *Oryza sativa* subspecies *indica*, which was developed in South Asia, and *Oryza sativa* subspecies *japonica*, which was developed in China.

CLASSICAL PLANT BREEDING

Classical plant breeding uses deliberate interbreeding (crossing) of closely or distantly related individuals to produce new crop varieties or lines with desirable properties. Plants are crossbred to introduce traits/genes from one variety or line into a new genetic background. For example, a mildew-resistant pea may be crossed with a high-yielding but susceptible pea, the goal of the cross being to introduce mildew resistance without losing the high-yield characteristics. Progeny from the cross would then be crossed with the high-yielding parent to ensure that the progeny were most like the high-yielding parent, (backcrossing). The progeny from that cross would then be tested for yield and mildew resistance and high-yielding resistant plants would be further developed. Plants may also be crossed with themselves to produce inbred varieties for breeding.

Classical breeding relies largely on homologous recombination between chromosomes to generate genetic diversity. The classical plant breeder may also makes use of a number of *in vitro* techniques such as protoplast fusion, embryo rescue or mutagenesis to generate diversity and produce hybrid plants that would not exist in nature.

Traits that breeders have tried to incorporate into crop plants in the last 100 years include:

- Increased quality and yield of the crop
- Increased tolerance of environmental pressures (salinity, extreme temperature, drought)
- Resistance to viruses, fungi and bacteria
- Increased tolerance to insect pests
- Increased tolerance of herbicides

BEFORE WORLD WAR II

Intraspecific hybridization within a plant species was demonstrated by Charles Darwin and Gregor Mendel, and was further developed by geneticists and plant breeders. In the United Kingdom in the 1880s, it was the pioneering work of Gartons Agricultural Plant Breeders. In the early 20th century, plant breeders realized that Mendel's findings on the non-random nature of inheritance could be applied to seedling populations produced through deliberate pollinations to predict the frequencies of different types.

From 1904 to World War II in Italy Nazareno Strampelli created a number of wheat hybrids. His work allowed Italy to increase hugely crop production during the so called "Battle for Grain" (1925–1940) and some varieties was exported in foreign countries, as Argentina, Mexico, China and others. After the war, the work of Strampelli was quickly forgotten, but thanks to the hybrids he created, Norman Borlaug was able to move the very first steps of the Green Revolution.

In 1908, George Harrison Shull described heterosis, also known as hybrid vigour. Heterosis describes the tendency of the progeny of a specific cross to outperform both parents. The detection of the usefulness of heterosis for plant breeding has led to the development of inbred lines that reveal a heterotic yield advantage when they are crossed. Maize was the first species where heterosis was widely used to produce hybrids.

By the 1920s, statistical methods were developed to analyze gene action and distinguish heritable variation from variation caused by environment. In 1933, another important breeding technique, cytoplasmic male sterility (CMS), developed in maize, was described by Marcus Morton Rhoades. CMS is a maternally inherited trait that makes the plant produce sterile pollen. This enables the production of hybrids without the need for labor intensive detasseling. These early breeding techniques resulted in large yield increase in the United States in the early 20th century. Similar yield increases were not produced elsewhere until after World War II, the Green Revolution increased crop production in the developing world in the 1960s.

AFTER WORLD WAR II

Following World War II a number of techniques were developed that allowed plant breeders to hybridize distantly related species, and artificially induce genetic diversity. When distantly related species are crossed, plant breeders make use of a number of plant tissue culture techniques to produce progeny from otherwise fruitless mating. Interspecific and intergeneric hybrids are produced from a cross of related species or genera that do not normally sexually reproduce with each other.

These crosses are referred to as *Wide crosses*. For example, the cereal triticale is a wheat and rye hybrid. The cells in the plants derived from the first generation created from the cross contained an uneven number of chromosomes and as result was sterile. The cell division inhibitor colchicine was used to double the number of chromosomes in the cell and thus allow the production of a fertile line. Failure to produce a hybrid may be due to pre- or post-fertilization incompatibility. If fertilization is possible between two species or genera, the hybrid embryo may abort before maturation. If this does occur the embryo resulting from an interspecific or intergeneric cross can sometimes be rescued and cultured to produce a whole plant. Such a method is referred to as *Embryo*

Rescue. This technique has been used to produce new rice for Africa, an interspecific cross of Asian rice *(Oryza sativa)* and African rice *(Oryza glaberrima)*. Hybrids may also be produced by a technique called protoplast fusion. In this case protoplasts are fused, usually in an electric field. Viable recombinants can be regenerated in culture.

Chemical mutagens like EMS and DMS, radiation and transposons are used to generate mutants with desirable traits to be bred with other cultivars - a process known as *Mutation Breeding*. Classical plant breeders also generate genetic diversity within a species by exploiting a process called somaclonal variation, which occurs in plants produced from tissue culture, particularly plants derived from callus. Induced polyploidy, and the addition or removal of chromosomes using a technique called chromosome engineering may also be used.

When a desirable trait has been bred into a species, a number of crosses to the favoured parent are made to make the new plant as similar to the favoured parent as possible. Returning to the example of the mildew resistant pea being crossed with a high-yielding but susceptible pea, to make the mildew resistant progeny of the cross most like the high-yielding parent, the progeny will be crossed back to that parent for several generations. This process removes most of the genetic contribution of the mildew resistant parent. Classical breeding is therefore a cyclical process. With classical breeding techniques, the breeder does not know exactly what genes have been introduced to the new cultivars. Some scientists therefore argue that plants produced by classical breeding methods should undergo the same safety testing regime as genetically modified plants. There have been instances where plants bred using classical techniques have been unsuitable for human consumption, for example the poison solanine was unintentionally increased to unacceptable levels in certain varieties of potato through plant breeding. New potato varieties are often screened for solanine levels before reaching the marketplace.

MODERN PLANT BREEDING

Modern plant breeding uses techniques of molecular biology to select, or in the case of genetic modification, to insert, desirable traits into plants.

STEPS OF PLANT BREEDING

The following are the major activities of plant breeding;

- Creation of variation
- Selection
- Evaluation
- Release
- Multiplication
- Distribution of the new variety

MARKER ASSISTED SELECTION

Sometimes many different genes can influence a desirable trait in plant breeding. The use of tools such as molecular markers or DNA fingerprinting can map thousands of genes. This allows plant breeders to screen large populations of plants for those that possess the trait of interest. The screening is based on the presence or absence of a certain gene as determined by laboratory procedures, rather than on the visual identification of the expressed trait in the plant.

REVERSE BREEDING AND DOUBLED HAPLOIDS (DH)

A method for efficiently producing homozygous plants from a heterozygous starting plant, which has all desirable traits. This starting plant is induced to produce doubled haploid from haploid cells, and later on creating homozygous/ doubled haploid plants from those cells. While in natural offspring genetic recombination occurs and traits can be unlinked from each other, in doubled haploid cells and in the resulting DH plants recombination is no longer an issue. There, a recombination between two corresponding chromosomes does not lead to un-linkage of alleles or traits, since it just leads to recombination with its identical copy. Thus, traits on one chromosome stay linked.

Selecting those offspring having the desired set of chromosomes and crossing them will result in a final F1 hybrid plant, having exactly the same set of chromosomes, genes and traits as the starting hybrid plant. The homozygous parental lines can reconstitute the original heterozygous plant by crossing, if desired even in a large quantity. An individual heterozygous plant can be converted into a heterozygous variety (F1 hybrid) without the necessity of vegetative propagation but as the result of the cross of two homozygous/doubled haploid lines derived from the originally selected plant. patent

GENETIC MODIFICATION

Genetic modification of plants is achieved by adding a specific gene or genes to a plant, or by knocking down a gene with RNAi, to produce a desirable phenotype. The plants resulting from adding a gene are often referred to as transgenic plants. If for genetic modification genes of the species or of a crossable plant are used under control of their native promoter, then they are called cisgenic plants. Genetic modification can produce a plant with the desired trait or traits faster than classical breeding because the majority of the plant's genome is not altered. To genetically modify a plant, a genetic construct must be designed so that the gene to be added or removed will be expressed by the plant. To do this, a promoter to drive transcription and a termination sequence to stop transcription of the new gene, and the gene or genes of interest must be introduced to the plant. A marker for the selection of transformed plants is also included. In the laboratory, antibiotic resistance is a commonly used marker:

Plants that have been successfully transformed will grow on media containing antibiotics; plants that have not been transformed will die. In some instances markers for selection are removed by backcrossing with the parent plant prior to commercial release. The construct can be inserted in the plant genome by genetic recombination using the bacteria *Agrobacterium tumefaciens* or *A. rhizogenes*, or by direct methods like the gene gun or microinjection. Using plant viruses to insert genetic constructs into plants is also a possibility, but the technique is limited by the host range of the virus. For example, Cauliflower mosaic virus (CaMV) only infects cauliflower and related species. Another limitation of viral vectors is that the virus is not usually passed on the progeny, so every plant has to be inoculated.

The majority of commercially released transgenic plants are currently limited to plants that have introduced resistance to insect pests and herbicides. Insect resistance is achieved through incorporation of a gene from *Bacillus thuringiensis* (Bt) that encodes a protein that is toxic to some insects. For example, the cotton bollworm, a common cotton pest, feeds on Bt cotton it will ingest the toxin and die. Herbicides usually work by binding to certain plant enzymes and inhibiting their action. The enzymes that the herbicide inhibits are known as the herbicides *target site*. Herbicide resistance can be engineered into crops by expressing a version of *target site* protein that is not inhibited by the herbicide. This is the method used to produce glyphosate resistant crop plants. Genetic modification of plants that can produce pharmaceuticals (and industrial chemicals), sometimes called *pharmacrops*, is a rather radical new area of plant breeding.

PARTICIPATORY PLANT BREEDING

The development of agricultural science, with phenomenon like the Green Revolution arising, have left millions of farmers in developing countries, most of whom operate small farms under unstable and difficult growing conditions, in a precarious situation. The adoption of new plant varieties by this group has been hampered by the constraints of poverty and the international policies promoting an industrialized model of agriculture. Their response has been the creation of a novel and promising set of research methods collectively known as participatory plant breeding. Participatory means that farmers are more involved in the breeding process and breeding goals are defined by farmers instead of international seed companies with their large-scale breeding programmes. Farmers' groups and NGOs, for example, may wish to affirm local people's rights over genetic resources, produce seeds themselves, build farmers' technical expertise, or develop new products for niche markets, like organically grown food.

GENETIC BASIS FOR THE BACKCROSS METHOD

In the F_1, many loci will be heterozygous (those loci which had different gene alleles coming from the two parents). These loci will include the gene of interest as well as many other loci containing genes for other traits. With backcrossing, homozygosity for the alleles from the recurrent parent will increase at the same rate as is seen with the approach to homozygosity with selfing (one form of inbreeding). The formula for calculating this rate is given below:

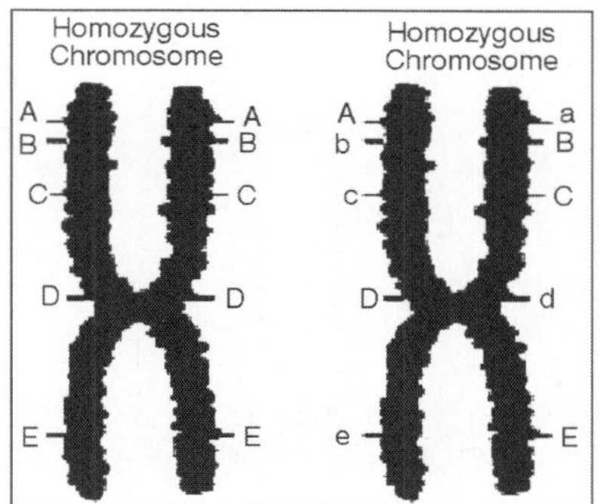

Fig. A Chromosome Homozygous for a Gene will have two Copies of the same Allele for that Gene. A Chromosome Heterozygous for a Gene will have two Different Alleles.

$$\left(\frac{(2^n - 1)}{2^n}\right)^m$$

where n = number of bc generations:
and m = number of loci.
Rr x Rr
Rr x rr $(2^1 - 1)^1$ or $((2 \times 1)-1)/2^1 = 1/2$ homozygosity
F_2: 1/4RR: 1/2Rr: 1/4rr
BC_1: 1/2Rr: 1/2rr
Similarly in the F_2 generation, 1/2 the plants are also homozygous (RR or rr).

Let's consider back the example of leaf rust resistance (RR, Rr). One of the goals of backcrossing is to remove the donor parent's genes (except the one of interest, RR; the others are nontarget genes) and increase the recurrent parent's genes except for the gene of disinterest (rr). The amount of remaining genetic information (the nontarget genes), on the average, from the nonrecurrent parent (donor parent) is reduced by 50% with each backcross.

The calculation for this data is below: (RR, Rr) Percentage of nontarget genes from donor parent = $(1/2)^{n+1}$

where n = number of backcross: However, the rate at which genes entering a cross from the donor parent (R = rust resistant) are eliminated during backcrossing will be influenced by linkage. Linkage is a term used to describe genes that do not independently assort (one of the requirements for Mendelian inheritance).

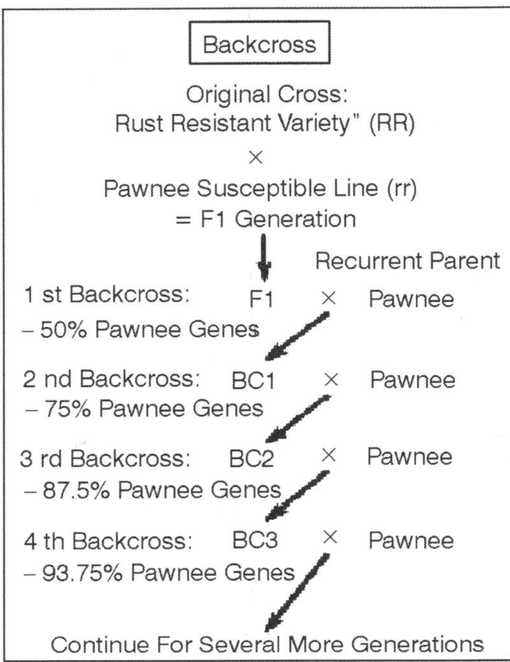

Fig. The Percentage of Recurrent Parent Genes Increases with Every Generation of Crossing.

Linked genes tend to stay together and unlinked genes independently assort. The physical basis for linkage is how close the genes are to each other on a chromosome. If genes are far apart or on different chromosomes, they are unlinked and independently assort. If they are near to each other on the chromosome they are more frequently passed on together than separately.

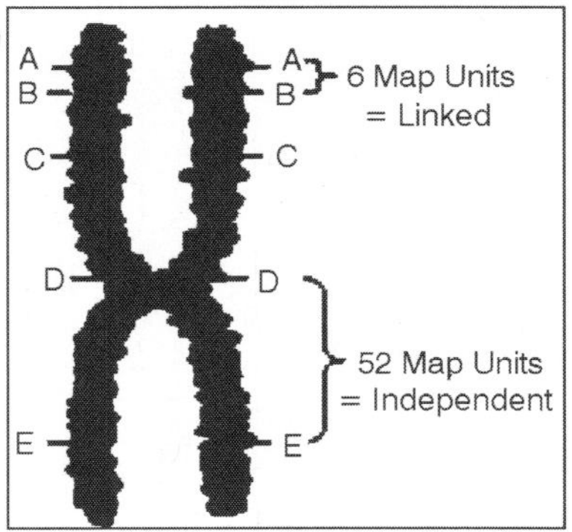

Fig. Genes Less than 50 map Units Apart are Considered Linked. The closer Genes are Located on a Chromosome, the more Likely they are to be Passed on to Offspring Together.

Linkage is measured by the recombination frequencies/map distance. Factors such as centromere location and chromosome structural abnormalities can reduce crossover events during meiosis. For example, if an undesirable allele d, for dwarfing is linked to R (rust resistant), and selection is only for R, d tends to be brought along in the F_1. However, when reintroducing R each backcross, the number of opportunities for crossing over between the R and d loci occur. Therefore, the probability of eliminating d is: $1 - (1-p)^n$

where, n = number of backcrosses.

p = recombination frequency between loci.

It should be noted that if d and R are close together (small map distance), it will be very hard to select R and is eliminated. Rarely do breeders know all the traits which are linked to the gene of interest in a backcrossing programme.

Requirements for a Successful Backcross

First is a satisfactory recurrent parent. A Good recurrent parents are those varieties or inbred lines that have maintained their importance for many years despite the release of numerous 'improved' varieties intended to replace them. A recurrent parent that has maintained its importance is one that remains high yielding in comparison to other lines over time. Remember that it often takes 5 to 7 years to develop a backcross line. In that 5 to 7 years, new cultivars are developed.

If the new cultivars with higher yield are better than the recurrent parent with the new trait, no one will grow the recurrent parent with the new trait. Thus the time and effort to back cross the new trait into the recurrent parent would be wasted except as a parent to develop other new cultivars. Secondlya gene is required hat is unaffected by environment or being in a new genetic background. High heritability of the character being transferred is important. Transfer is easiest when the character can be identified readily in hybrid populations by visual selection or by simple tests. There has been some success in dealing with quantitative traits through backcrossing; however, like all other methods it depends on the ability of the breeder to distinguish between genetic and environmental variability, and to select those individuals that are desirable for genetic reasons.

The general agricultural worth of the donor parent need not be of great concern. Selection of the donor, (nonrecurrent parent) is almost exclusively on the basis that it exhibits the character in a particularly intense form. This is important since often the presence of modifier genes in the new genetic background (recurrent parent) cause some intensity to be lost even though the most stringent selection has been practiced throughout the backcrossing programme.

A final absolute requirement is a sufficient number of backcrosses must be used to reconstitute the recurrent parent to a high degree. Recovery of the recurrent parent is a function of the number of backcrosses and the effectiveness of selection for recurrent parent type in the early generations. Usually 6 backcrosses with selection for type in the early generations has proved sufficient. However, in wide crosses and/or with undesirable linkages a greater number of backcrosses may be necessary.

Fig. Wheat Breeding in a Greenhouse. Wheat Heads are Bagged for Proper Pollination.

Remember that in practice the objective is to improve the recurrent parent by only one trait. It should be noted that since the recurrent parent is already a proven variety or line, it is usually not necessary to conduct extensive performance trials once satisfactory introduction of the desired character has been achieved. Previous data on the recurrent parent will indicate the benefits of the backcross line assuming the added gene or linked genes are not detrimental. The last consideration a breeder is aware of is the influence of environmental conditions on a backcross programme. Provided the expression of the character being transferred is sufficient for selection, backcrossing can be conducted in any environment. For example, several generations may be grown per year in the greenhouse.

Improvement of More Than One Character

Often in a variety development programme, the breeder wants to add more than one gene to a variety. This is known as gene stacking. It is possible to introduce several different characters in the course of a backcross programme. Usually the other characters being added have already been introduced into the recurrent parent in other backcross programmes.

For example, one breeder wanted to develop rust and bunt (another disease) resistant Baart 38 wheat. He did this by backcrossing rust resistance into Baart and by backcrossing bunt resistance into a second line of Baart. He then crossed the rust resistant Baart and the bunt resistant Baart and selected Baart that was both rust and bunt resistant. Currently, backcrossing is often used to develop improved gene pools or selection populations without trying to return completely to the recurrent parent's phenotype. In some cases, the variety has a good phenotype and the breeder wants to improve it without having a clearly defined trait (for example, yield). In order to meet this objective, the breeder can use different numbers of backcrosses (1 to 6) to add as much or as little of the recurrent parent as he/she thinks is necessary. A backcross one population (A x (A x B)) is preferred by many breeders to a simple three-way cross (A x (B x C)), which may have too much variation. In this case A, B, and C represent elite lines or varieties.

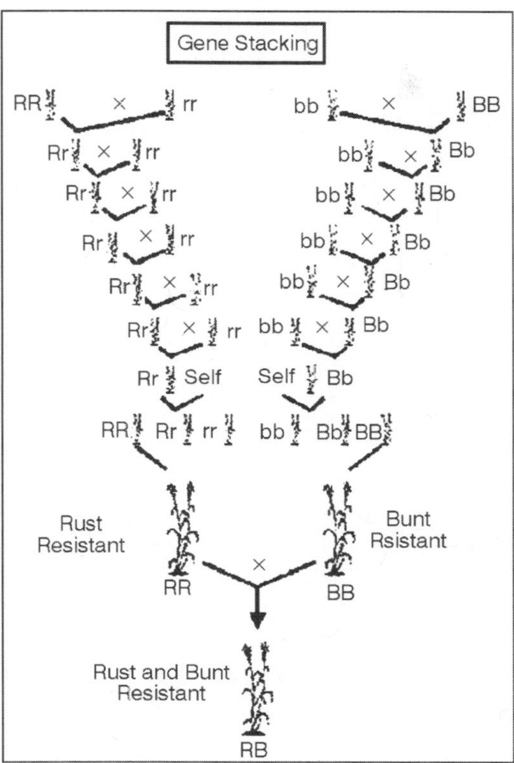

Fig. Using Breeding to Stack Multiple Traits into a Single Line.

Importance of back Cross Method

There are both strengths and weaknesses of the backcross breeding method. Properly executed, the backcross breeding methods allow all the desirable characteristics of the recurrent parent to be recovered with the possible exception for characters governed by genes tightly linked to the gene(s) being transferred. This may be considered a strength. The backcross method provides a certain and precise way of making predictable gains with little possibility that uncontrolled segregation will produce subtle weaknesses in the variety.

This same characteristic may also be considered a weakness. The method sets an upper limit which is equal to the recurrent parent's phenotype. This limit will often be lower than the progress that is possible with other lines when segregation is not rigidly controlled. This phenomenon is known as yield lag and is graphically depicted in Figure below. Series 1 shows the gradual per cent improvements reached over time. In contrast, Series 2 is put through a backcross programme at year 4, where its per cent improvement plateaus.

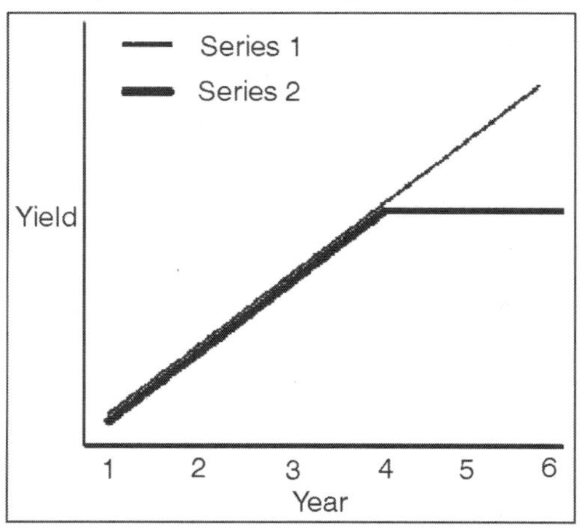

Fig. Series 1: Gradual Per cent Improvements in Yield Reached over time. Series 2: A line is put Through a Backcross Programme at year 4.

EUPLOIDY IN BREEDING

The number of chromosomes in a basic set is called the monoploid number (x). Organisms with multiples of the monoploid number of chromosomes are called euploid. Haploids and diploids, then, are both cases of normal euploidy. Euploid types that have more than two sets of chromosomes are called polyploid. The polyploid types are named triploid ($3x$), tetraploid ($4x$), pentaploid ($5x$), hexaploid ($6x$), and so forth. Polyploids arise naturally as spontaneous chromosomal mutations and, as such, they must be considered aberrations because they differ from the previous norm.

However, many species of plants and animals have clearly arisen through polyploidy, so evidently evolution can take advantage of polyploidy when it arises. It is worth noting that organisms with one chromosome set sometimes arise as variants of diploids; such variants are called monoploid ($1x$). In some species, monoploid stages are part of the regular life cycle, but other monoploids are spontaneous aberrations. The haploid number (n), which we have already used extensively, refers strictly to the number of chromosomes in gametes. In most plants with which we are familiar, the haploid number and monoploid number are the same.

Hence, n or x (or $2n$ or $2x$) can be used interchangeably. However, in certain plants, such as modern wheat, n and x are different. Wheat has 42 chromosomes, but careful study reveals that it is hexaploid, with six rather similar but not identical sets of seven chromosomes. Hence, $6x = 42$ and $x = 7$. However, the gametes of wheat contain 21 chromosomes, so $n = 21$ and $2n = 42$.

Monoploids

Male bees, wasps, and ants are monoploid. In the normal life cycles of these insects, males develop parthenogenetically—that is, they develop from unfertilized eggs. However, in most species, monoploid individuals are abnormal, arising in natural populations as rare aberrations. The germ cells of a monoploid cannot proceed through meiosis normally, because the chromosomes have no pairing partners. Thus, monoploids are characteristically sterile. (Male bees, wasps, and ants bypass meiosis in forming gametes; here, mitosis produces the gametes.) If a monoploid cell does undergo meiosis, the single chromosomes segregate randomly, and the probability of all chromosomes going to one pole is $(1/2)^{x1}$ where x is the number of chromosomes. This formula estimates the frequency of viable (whole-set) gametes, which is a small number if x is large.

Monoploids play an important role in modern approaches to plant breeding. Diploidy is an inherent nuisance when breeders want to induce and select new gene mutations that are favourable and to find new combinations of favourable alleles at different loci. New recessive mutations must be made homozygous before they can be expressed, and favourable allelic combinations in heterozygotes are broken up by meiosis. Monoploids provide a way around some of these problems. In some plant species, monoploids can be artificially derived from the products of meiosis in a plant's anthers. A cell destined to become a pollen grain can instead be induced by cold treatment to grow into an embryoid, a small dividing mass of cells. The embryoid can be grown on agar to form a monoploid plantlet, which can then be potted in soil and allowed to mature.

Plant monoploids can be exploited in several ways. In one, they are first examined for favourable traits or allelic combinations, which may arise from heterozygosity already present in the parent or induced in the parent by mutagens. The monoploid can then be subjected to chromosome doubling to achieve a completely homozygous diploid with a normal meiosis, capable of providing seed. How is this achieved? Quite simply, by the application of a compound called colchicine to meristematic tissue. Colchicine—an alkaloid drug extracted from the autumn crocus—inhibits the formation of the mitotic spindle, so cells with two chromosome sets are produced. These cells may proliferate to form a sector of diploid tissue that can be identified cytologically. Another way in which the monoploid may be used is to treat its cells basically like a population of haploid organisms in a mutagenesis-and-selection procedure. A population of cells is isolated, their walls are removed by enzymatic treatment, and they are treated with mutagen. They are then plated on a medium that selects for some desirable phenotype. This approach has been used to select for resistance to toxic compounds produced by one of the plant's parasites and to select for resistance to herbicides being used by farmers to kill weeds.

Resistant plantlets eventually grow into haploid plants, which can then be doubled (with the use of colchicine) into a pure-breeding, diploid, resistant type. These powerful techniques can circumvent the normally slow process of meiosis-based plant breeding. The techniques have been successfully applied to several important crop plants, such as soybeans and tobacco. The anther technique for producing monoploids does not work in all organisms or in all genotypes of an organism. Another useful technique has been developed in barley, an important crop plant. Diploid barley, *Hordeum vulgare,* can be fertilized by pollen from a diploid wild relative called *Hordeum bulbosum.*

This fertilization results in zygotes with one chromosome set from each parental species. In the ensuing somatic cell divisions, however, the chromosomes of *H. bulbosum* are eliminated from the zygote, whereas all the chromosomes of *H. vulgare* are retained, resulting in a haploid embryo. (The haploidization appears to be caused by a genetic incompatibility between the chromosomes of the different species.) The chromosomes of the resulting haploids can be doubled with colchicine. This approach has led to the rapid production and widespread planting of several new barley varieties, and it is being used successfully in other species too.

Polyploids

In the realm of polyploids, we must distinguish between autopolyploids, which are composed of multiple sets from within one species, and allopolyploids, which are composed of sets from different species. Allopolyploids form only between closely related species; however, the different chromosome sets are homeologous (only partly homologous)—not fully homologous, as they are in autopolyploids.

Triploids

Triploids are usually autopolyploids. They arise spontaneously in nature or are constructed by geneticists from the cross of a $4x$ (tetraploid) and a $2x$ (diploid). The $2x$ and the x gametes unite to form a $3x$ triploid. Triploids are characteristically sterile. The problem, like that of monoploids, lies in pairing at meiosis. Synapsis, or true pairing, can take place only between two chromosomes, but one chromosome can pair with one partner along part of its length and with another along the remainder, which gives rise to an association of three chromosomes. Paired chromosomes of the type found in diploids are called bivalents. Associations of three chromosomes are called trivalents, and unpaired chromosomes are called univalents. Hence in triploids there are two pairing possibilities, resulting either in a trivalent or in a bivalent plus a univalent.

Paired centromeres segregate to opposite poles, but unpaired centromeres pass to either pole randomly. If all the single chromosomes pass to the *same* pole and simultaneously the other two chromosomes pass to the opposite pole, then the gametes formed will be haploid and diploid. The probability of this type of meiosis will be $(1/2)^{x1}$, and this proportion is likely to be low. All other possibilities will give gametes with chromosome numbers intermediate between the haploid and diploid number; such genomes are aneuploid—"not euploid." It is likely that these aneuploid gametes will not lead to viable progeny; in fact, this category is responsible for the almost complete lack of fertility of triploids.

The problem is one of genome imbalance. For most organisms, the euploid chromosome set is a finely tuned set of genes in relative proportions that seem to be functionally significant. Multiples of this set are tolerated because there is no change in the relative proportions of genes. However, the addition of one or more extra chromosomes is nearly always deleterious because the proportions of genes in those extra chromosomes are altered. Although the action of some genes can be regulated to compensate for extra gene "dosage," the overall effect of the extra genetic material seems too great to be overcome by gene regulation. The deleterious effect can be expressed at the level of gametes, making them nonfunctional, or at the level of the zygote, resulting in lethality, sterility, or lowered fitness. In triploids, it is possible that some haploid or diploid gametes will form, and some may unite to form a euploid zygote, but the likelihood of this possibility is inherently low.

Consider bananas. The bananas that are widely available commercially are triploids with 11 chromosomes in each set ($3x = 33$). The probability of a meiosis in which all univalents pass to the same pole is $(1/2)^{x1}$, or $(1/2)^{10} = 1/1024$, so bananas are effectively sterile. The most obvious expression of the sterility of bananas is that there are no seeds in the fruit that we eat. Another example of the commercial exploitation of triploidy in plants is the production of triploid watermelons. For the same reasons that bananas are seedless, triploid watermelons are seedless, a phenotype favoured by some for its convenience.

AutotetrAploids

Autotetraploids arise naturally by the spontaneous accidental doubling of a $2x$ genome to a $4x$ genome, and autotetraploidy can be induced artificially through the use of colchicine. Autotetraploid plants are advantageous as commercial crops because, in plants, the larger number of chromosome sets often leads to increased size. Cell size, fruit size, flower size, stomata size, and so forth, can be larger in the polyploid.

Epidermal leaf cells of tobacco plants, showing an increase in cell size, particularly evident in stoma size, with an increase in autopolyploidy. (a) Diploid, (b) tetraploid, (c) octoploid. Because 4 is an even number, autotetraploids can have a regular meiosis, although this is by no means always the case. The crucial factor is how the four homologous chromosomes, one from each of the four sets, pair and segregate.

There are several possibilities, as shown in Figure. Pairings between four chromosomes are called quadrivalents. In tetraploids, the two-bivalent and the quadrivalent pairing modes tend to be most regular in segregation, but even here there is no guarantee of a 2:2 segregation. If all chromosome sets segregate 2:2 as they do in some species, then the gametes will be functional and a formal genetic analysis can be developed for such autotetraploids.

Let's consider the genetics of a fertile tetraploid. We can consider an experiment in which colchicine is used to double the chromosomes of an *A/a* plant to form an *A/A/a/a* auto-tetraploid, which we will assume shows 2:2 segregation. We now have a further concern because polyploids such as tetraploids give different phenotypic ratios in their progeny, depending on whether the locus in question is tightly linked to the centromere. First, we consider a centromere-linked gene. The three possible pairing and segregation patterns are presented in Figure; these patterns occur by chance and with equal frequency. As shows, the 2*x* gametes produced are *A/a*, *A/A*, or *a/a*, in a ratio of 8:2:2, or 4:1:1. If such a plant is selfed, the probability of an *a/a/a/a* phenotype in the offspring is 1/6 × 1/6 =1/36. In other words, a 35:1 phenotypic ratio of *A/ //:a/a/a/a* will be observed if *A* is fully dominant over three *a* alleles.

If, in the same kind of plant, a genetic locus having the alleles *B* and *b* is very far removed from the centromere, crossing-over must be considered. This example forces us to think in terms of chromatids instead of chromosomes; there are four *B* chromatids and four *b* chromatids. Because the number of crossovers in such a long region will be large, the genes will become effectively unlinked from their original centromeres. The packaging of genes two at a time into gametes is very much like grabbing two balls, at random from a bag of eight balls: four of one kind and four of another. The probability of picking two *b* genes is then So, in a selfing, the probability of a *b/b/b/b* phenotype is 2/4 × 3/ 14 = 9/196, which is approximately 1/22. Hence, there will be a 21:1 phenotypic ratio of *B///:b/b/b/b*. For genetic loci of intermediate position, intermediate ratios will result.

Allopolyploids

G. Karpechenko synthesized the "classic" allopolyploid in 1928. He wanted to make a fertile hybrid that would have the leaves of the cabbage *(Brassica)* and the roots of the radish *(Raphanus)*. Each of these species has 18 chromosomes, and they are related closely enough to allow intercrossing. A viable hybrid progeny individual was produced from seed. However, this hybrid was functionally sterile because the nine chromosomes from the cabbage parent were different enough from the radish chromosomes that pairs did not synapse and disjoin normally. However, one day a few seeds were in fact produced by this (almost) sterile hybrid. On planting, these seeds produced fertile individuals with 36 chromosomes. All these individuals were allopolyploids. They had apparently been derived from spontaneous, accidental chromosome doubling to $2n_1 + 2n_2$ in the sterile hybrid, presumably in tissue that eventually became germinal and underwent meiosis. Thus, in $2n_1 + 2n_2$ tissue, there is a pairing partner for each chromosome and balanced gametes of the type $n_1 + n_2$ are produced.

These gametes fuse to give $2n_1 + 2n_2$ allopolyploid progeny, which also are fertile. This kind of allopolyploid is sometimes called an amphidiploid, which means "doubled diploid". (Unfortunately for Karpechenko, his amphidiploid had the roots of a cabbage and the leaves of a radish.) When the allopolyploid was crossed with either parental species, sterile offspring resulted. The offspring of the cross with radish were $2n_1 + n_2$ constituted from an $n_1 + n_2$ gamete from the allopolyploid and an n_1 gamete from the radish. The n_2 chromosomes had no pairing partners, so sterility resulted. Consequently, Karpechenko had effectively created a new species, with no possibility of gene exchange with its parents. He called his new species Raphanobrassica.

Today, allopolyploids are routinely synthesized in plant breeding. Instead of waiting for spontaneous doubling to occur in the sterile hybrid, the plant breeder adds colchicine to induce doubling. The goal of the breeder is to combine some of the useful features of both parental species into one type. This kind of endeavor is very unpredictable, as Karpechenko learned. In fact, only one synthetic amphidiploid has ever been widely used. This amphidiploid is *Triticale,* an amphidiploid between wheat. In nature, allopolyploidy seems to have been a major force in speciation of plants. There are many different examples. One particularly satisfying one is shown by the genus *Brassica,* as illustrated in. Here three different parent species have hybridized in all possible pair combinations to form new amphidiploid species. A particularly interesting natural allopolyploid is bread wheat, *Triticum aestivum*($2n = 6x = 42$). By studying various wild relatives, geneticists have reconstructed a probable evolutionary history of bread wheat. In a bread wheat meiosis, there are always 21 pairs of chromosomes. Furthermore, it has been possible to establish that any given chromosome has only one specific pairing partner (homologous pairing)—not five other potential partners (homeologous pairing). The suppression of such homeologous pairing (which would make the species more unstable) is maintained by an allele, *Ph*, on the long arm of chromosome 5 of the B set. Thus, *Ph* ensures a diploidlike meiotic behaviour for this hexaploid species. Without *Ph*, bread wheat could probably never have arisen. It is interesting to speculate about whether Western civilization could have begun or progressed without this species—in other words, without the *Ph* mutation.

SCOPE OF PLANT BREEDING

IMPROVING CROP PERFORMANCE

The dramatic gains in agricultural productivity seen in the second half of the last century are often linked uniquely to increased mechanisation and the widespread introduction of fertilisers and agrochemicals.This 'Green Revolution' would not have been possible without the huge genetic improvements made to the crops. Plant breeding has contributed around half of the threefold increase in UK wheat yields recorded from 1947 to 1992 and yields have continued to improve since then. Plant breeding can directly improve the performance of crops in different ways.

Productive Yield

Developing crop varieties, which convert more of their biomass into productive yield, is the single biggest contributor to improved crop output. The introduction of shorter-strawed cereals is a striking example of how this has been achieved, by transforming more of the crop's productive energy into valuable grain.

Physical Characteristics

Changing a crop's physical structure can also contribute to increased yields. For example, the development of semi-leafless varieties of field peas helped to boost intrinsic yield. It also stimulated the growth of stabilising tendrils, which significantly improved the crop's standing ability so reducing crop losses at harvest.

Disease Resistance

Genetic resistance to disease enables crops to realise their yield potential – it can also mean reduced use of agrochemicals. Plant breeding has significantly improved the genetic resistance of crops against the threat of viral and fungal infection. Key examples include resistance to blight in potatoes, rhizomania in sugar beet, and barley yellow mosaic virus in cereals. The challenge of breeding resistant varieties is constant because new strains of disease develop naturally.

Timing of Maturity

Plant breeding technology has brought major improvements in the uniformity with which crops ripen ready for harvest. This not only reduces potential crop losses at harvest (as in the case of pod shatter in oilseed rape), but has also improved growers' ability to mechanise harvesting operations. In the field vegetable sector, for example, the labour intensive (and unpleasant!) task of picking crops such as cauliflower, broccoli and Brussels sprouts has been transformed by the development of improved varieties.

Other Agronomic Factors

By improving crops' ability to cope with a range of other agronomic pressures, advances in plant breeding continue to underpin progress in agricultural productivity.

They include:

- Genetic resistance to pests, such as nematode resistance in potatoes

- Shorter crop life-cycle, to expand a crop's geographical growing areas
- Stress tolerance, such as frostresistance in field vegetables, to extend the seasonal availability of home-grown fresh produce.

Improving Crop Quality

Alongside progress in keeping Britain's farmers competitive through improved productivity, developments in plant breeding have also brought significant gains in crop quality. Gone are the days when Europe encouraged farmers to concentrate solely on maximising output, assured that their crops would find support under the Common Agricultural Policy. Today, as production-based support and barriers to international trade continue to be reduced, Britain's growers must compete with the best in the world, on quality and cost of production. Consumer expectations of food are also more exacting. This is reflected in tighter quality specifications along the food chain. Plant breeders have responded with a continuous stream of new varieties, tailored to the needs of specific end-markets.

Cereals

Britain's historical dependence on North American imports of wheat for bread making has been reversed over the past 40 years thanks to improvements in baking technology and the development of high protein, hard milling varieties of wheat. Until the mid-1960s, imported wheat typically accounted for up to 65% of our bread making requirements. Today as much as 90% of our bread is produced from homegrown varieties. In barley, improvements in malting quality have underpinned a fourfold increase in the volume of beer brewed per tonne, from around 2,000 liters in 1950 to nearly 8,000 liters today. The quality of malting barley varieties grown in Britain is internationally recognized, with export markets from the near Continent to the Far East.

Oilseed Rape

Oilseed rape has become a major homegrown source of vegetable oil and animal feed within the past 30 years. The rapid expansion of this important break crop from 50,000 hectares in the mid-1970s to around 500,000 hectares today can be linked to two major breakthroughs in plant breeding. Varieties for food use were first developed in 1970. These contain reduced levels of erucic acid in the oil, making it more suitable for human consumption. This was followed in the late 1980s by 'doublelow' varieties, which also offered reduced levels of glucosinolates in the residual meal to improve quality for animal feed.

Potatoes

In the potato sector, plant breeders have responded to increasing consumer demands for quality and choice. Fresh produce counters now boast a bigger range than ever, from traditional main-crop varieties through to baby new, baking and salad potatoes. Breeding programmes have also targeted processing and catering markets, with specific varieties used to produce crisps, instant mash and chips.

Sugar Beet

Higher root yield and increased sugar content have combined to double sugar production per hectare in the past 50 years. The introduction of monogerm seed, allowing seeds to be planted individually rather than in clusters, ensures genetic improvements are fully realised in the field.

Pulses

As interest in home-grown sources of protein has increased, breeders have developed varieties of peas and beans for specialist markets – freezing and canning for human consumption, flaking for pet food, and tannin-free varieties for animal feed.

Creating a New Variety

The creation of each new variety is a complex, costly and skilled operation. It is also painstakingly slow – today's breeding programmes are already looking ten years ahead to the needs of farmers, consumers and the environment at the end of the next decade and beyond. Techniques vary between crop species, but the scientific principles of plant breeding remain unchanged from Mendel's first discovery that selected parent plants can be cross-pollinated to combine desired characteristics.

Genes – units of hereditary material that are transferred from one generation to the next, determine plant characteristics. Since each plant contains many thousands of genes, and the breeder is seeking to combine a range of traits in one plant, such as high yield, quality and resistance to disease, developing a successful variety is an extremely lengthy process – up to 12 years in the case of cereals, even longer for potatoes. Plant breeding has been compared to playing a fruit machine – not with three reels, but several hundred. The skill of the plant breeder lies in improving his chances of hitting the jackpot by combining all the desired characteristics in the same variety.

SPEEDING-UP PLANT-BREEDING WORK

Plant-breeding work involving crosses generally takes a great number of generations before the superior new plant has been obtained. In the routine work of a large plant-breeding station so many breeding projects arc under way simultaneously that it matters little how much time elapses between the initial cross and the final purification of a promising commercial variety. Every year there are many projects being started, several are under way, and a few arc in their last stages. It often pays to cut down the time required for work of this kind. One of the most promising means of doing this is to grow more than one generation of plants in one year. This usually means that one of those generations of plants is grown in a season in which the plant is not ordinarily grown, and that for this reason a correct evaluation of the quality of the individual plants is out of the question. In Europe or in America growing an extra generation on the spot means growing plants in the greenhouse. To grow cereals or beans in winter in greenhouse conditions not only requires heat; we must also have the means of regulating the length of day. The simplest method is to use large electric lights to furnish both the heat and the light, to lengthen the day by those means, and to let the temperature go down during the dark part of the night.

Under those conditions cereals and beans will flower and set seed very well in the winter months, The electricity required per square yard is roughly that used by lamps that require a kilowatt. It is clear that under such conditions the winter generation will not show the normal qualities of the plants in that group which are grown in summer-time in the field, but there are many cases in which this is not of much importance. If we want to proceed by the method of making some hybrids and then growing a second and third inbred generation from this stock, we can do our crossing in the field, grow our hybrids in the winter, and then grow another crop of second-generation plants in normal field conditions. When we are using the method of mating back hybrid stock to some pure kind of plants we want to improve it matters but little if we cannot very well select the best plants during some of the generations involved. It is evident that we must grow a generation under normal field conditions whenever we want to make our selections.

In recent years a few seed-firms have hit upon the perfect scheme of growing two generations of annual plants. This does not involve any greenhouse generations. They work in two different localities with very similar climate, but so chosen that one of those localities is situated in the Southern Hemisphere. There are regions of Chili where Californian firms can find conditions of plant-growth that are identical with conditions at home, with the exception that the growing season comes during the Californian winter. In those circumstances it is quite possible to grow two absolutely normal generations during one year. Our present greatly improved methods of communication make such schemes perfectly feasible. With many tropical plants one generation yearly is grown of plants that need special conditions of either a wet or a dry season. Here it is easy enough to interpolate an additional generation, by looking for a suitable spot, or by providing the necessary irrigation.

Sometimes it takes a very long time to bring the plants to a condition where we can evaluate them. A good example is that of the fruit trees. It takes many years for an apple seedling or a seed-grown cherry to come into bearing. Here we can save a great deal of time by budding or top-grafting parts of the seedlings upon mature trees in full bearing. By this method we may save more than half the number of years normally required for bringing the seedlings to maturity. In this connection we might also treat of those cases where certain seeds take a very long time to germinate. Storing the seed in suitable conditions of moisture, and especially of temperature, will speed-up germination in such plants as gooseberries and celery. In stone-fruits it has been found possible to crack open the stone, to extract the kernel and grow the seedlings under aseptic or even under antiseptic conditions (thymol solution).

SHOWS AND SHOWING

The exhibiting of plants and seed at the shows is a very curious thing. It pays extremely well for a seed-firm to employ some gardeners who know just exactly when and' how to get enough plants full of luscious beans or covered with flowers on the exact day for each show.

Florists and lovers of gardening see the quality of such plants at the shows, and many people who admire them there try in their turn to win the coveted prizes at their local shows the next year. This show game is certainly very nice and pleasant; it gives the gardeners something to strive after; it may even help to keep some of the lads in the country. On the other hand, there is very, very little connection between the qualities judged at the shows and actual value of the material shown. The few selected mangolds, or the mammoth pumpkins, or the bundle of wheat- ears, or even of the maize-ear contests of the American agricultural shows.

All along the line, in horticulture as well as in agriculture, the selected samples shown give no indication of the value of the seed or plant stock the producers have for sale. Of course we can see that those few mangolds have the ideal shape and size, but the seed-firm must give us some guarantee about the percentage of mangolds that will grow to this size and shape from the seed they sell, just as the owner of a prize-winning bull must give some guarantee of this animal's ability to give profitable daughters, in addition to the blue ribbon awarded to the beautifully coloured and shaped prize-winner. In agriculture this is beginning to be so well realised that the classes for beautiful corn cobs or for the largest mangolds are becoming very rare indeed, even if we still see stock-judging in dairy cows and similar farces. In the horticultural shows much interest is still shown in regard to flowers and vegetables.

It seems evident that when we strive for beauty, beauty contests for roses and begonias, for asparagus and delphiniums and dahlias are indicated. The show is a shop-window. The seller must show if his competitors show, if only to keep his name before the public. To the public the show is only just a pleasant occasion for a half-day outing. To the buyer the show is much nicer to see than a description in a catalogue.

Long lists of novelties are bought at the shows by people with an experimental turn of mind. We see the representatives at the shows kept busy noting down long lists of orders. Showing certainly pays the seller. A very much more sensible system of showing a great many varieties is that of the demonstration gardens and experimental stations, both stations in which the agricultural authorities try to compare promising new and good old varieties with the aid of objective standards, and such gardens as show-plots arranged by rose-lovers or by societies of growers of gladioli or dahlias. Here the varieties can be judged throughout the growing season in favourable conditions.

THE ANNUAL SELF-FERTILIZING CROPS

The number of plant species now being cultivated by man is really enormous, while only relatively few animal species are kept under domestication. It is possible for the author of a book on animal breeding to take those species almost one by one, and to give the specialist breeders a few hints over and above those which are contained in the general book. This would be quite out of the question in a book on plant breeding. One way out of the difficulty would be to give one example of each of the three groups—self-fertilized plants, cross-fertilizers and plants that are usually propagated by vegetative means.

These plants in two sections— grains and legumes—giving some hints to the breeders of wheat, barley, oats, sorghum, rice, etc., and going into some details about the breeding of beans, soybeans, cow-peas, ground-nuts, peas, etc., separately. But, after all, the self-fertilizing annuals have so much in common from the point of view of plant-breeding methods. Self-fertilization causes isolation. Each plant, as far as reproduction is concerned, is as self-contained and isolated in a crowded field as it would be if grown by itself in a greenhouse in a city. The inbreeding caused by self-fertilization has several results.

It makes all heritable variability disappear in every line, it makes every group consist of parallel lines, in which each plant descends from only one parent plant, so that the usual complicated network of ancestry is simplified into a system of isolated parallel lines. Where inherited variability exists, such as happens after a deliberate artificial cross or when accidental cross-fertilization occurs, it will automatically disappear. That is to say, it will disappear in so far as each separate line is concerned, for Mendelian segregation will make the lines, descending from any hybrid heterozygous for a great many genes, diverse in their hereditary make-up.

The result is that a field of self-fertilizing annuals that has not been grown from just one pure line by a plant breeder always consists of a mixture of a number of genetically different lines. Even if accidental cross-fertilization or irregularities in crossing-over (mutation) are very rare, they do occur, and cause this state of things. This genetic variability, however, is wholly different from what we find in other plants and animals. We are dealing with mixtures of pure lines, and almost every plant we take from any field is pure, homozygous for all its genes, and will, if its seeds are sown, give a very homogeneous descendance. From all this it follows that enormous progress can often be made in all self-fertilized plants by simply picking out a great number of promising-looking individual plants and comparing their descendance.

Many valuable widely grown wheats, tobaccos, peanuts and beans have been just found. They "just growed ", like Topsy. The first step in plant-breeding in the self-fertilizing plants must always be to look for those ready-made pure lines. The next is to find whether a mixture of two of the very best pure lines can be discovered which will give us still higher production (10 per cent, higher production from a mixture of two strains as compared with that of the best one is quite common). We have then to produce some more genetic variability by deliberate cross-breeding. This generally means that we try to combine the good qualities of two different strains in one new one.

Here there are two wholly different schools of thought, and two wholly different methods. One of them is to analyse second-generation plants, and to find those individuals that do combine, let us say, the disease resistance of line A with the large quantity of big seeds of line B. This is a method in- vented by the experimental geneticists who are trying to combine practical plant breeding with purely scientific gene- analysis. As a theoretical geneticist, interested in genes and their action. It is great fun playing with a hundred different hybrid lots in one species, growing a few dozen second-generation plants in each lot and tabulating the results. But this is neither sound genetics nor sound plant breeding. We are dealing with so many unanalysed and unanalysable genes which all affect production and quality at the same time, that the chance of finding something really worth while in a few hundred plants is simply not good enough.

The other method is based upon the realization that hundreds of genes are commonly involved, and upon the fact that purification—reduction of the potential variability—is rapid and automatic in this material. It consists of growing the greatest possible number of seeds from first-generation hybrid plants. We can then proceed in two different ways. One is an analytical method, which means that we are seeking for the very best-looking plants (for instance, by weighing their seeds), and progeny testing those plants by sowing a large number of seeds from each, in order to find the most profitable families. The other method is what Baur and Nilsson-Ehle called a "Ramsch" method. This consists simply of growing the descendance of a hybrid in a vast mixture, refraining from any analysis, and continuing this hybrid lot as a miscellaneous collection for five or six generations, sowing as much of the mixed seed of each harvest as we have room for. If we do this, natural selection will weed out most of the unproductive and undesirable types, such as those plants t1 at ripen after harvesting time, all weaklings and all plants susceptible to diseases and pests. After five or six years that lot will consist of a mixture of pure lines, every line different from every other, with the successful lines represented by more individuals than the unsuccessful ones. After this we will treat the mixture just as if we were dealing with any other mixture, growing separate rows of beds each from one likely-looking plant. This is a very sound method, which saves a great deal of un- necessary labour and time. It is extremely difficult to find just how much better this method is as compared with that of analysis in every generation, starting from the second generation. An analytical method could be made to work very well, provided we did all we could to make the number of plants of the second and third generations exceedingly large (in the cereals we can get almost as many seeds as in tobacco by splitting up the hybrid plant repeatedly before it starts to grow any stalks).

In actual practice there are only a few crosses that really produce anything worth while, and the general practice seems to be to make a great many attempts and grow a large number of first-generation hybrids. This renders it imperative to cut down on the number of plants per experiment. It is a vicious circle. This is all very wrong and wasteful. One cross every year and work with it, both analytically and by means of a "Ramsch ", doing the final selection after six years for one lot in every season, than waste my time castrating flowers and producing numerous hybrids that would have no really good chance of showing what they could do. In all the self-fertilizing annuals we must be continually on the look-out for qualities that may be of special merit. Mostly of adaptations to special conditions. In wheaa on irrigated ground, resistance to drowning may be as important as disease resistance or quality of the grain. Conversely, rice could be grown much more extensively if we would take the trouble to do some plant-breeding work on dry-land rices, which abound in the mountain villages of all the tropical Asian islands.

In ground-nuts we should look for lines that are specially adapted to mechanical harvesting, as well as for lines adapted to the production of a first-quality product in the hands of tens of thousands of small cultivators. In all the plants of this group it is possible not only to improve the yield and the quality, but also to extend the area in which profitable production is possible. Tobacco, rice, soybeans and cow-peas are just as important in this respect as wheat, barley and oats. Plant breeding with self-fertilizing plants is comparatively easy. It is even possible for the real amateur plant breeder to do some extremely interesting and useful work with some of the less well-known plants of this group—let us say with beans or cow-peas. Such work needs room, but very little special technical knowledge or talent for higher arithmetic. Frequently a few plants will stand out from an otherwise quite even field of barley, beans or sorghum.

PLANT BREEDING STRATEGIES AND PRIORITIES

Plant breeding has been recognised as the major tool for green revolution and has been instrumental in boosting crop production in both developed and developing countries. High yielding varieties of rice and wheat were at the heart of the spectacular increase in food production in Asia and elsewhere. In CWANA region, modern varieties contributed up to 43 per cent of barley at production during the first phase of adoption. A recently published study estimating the value of the CGIAR's activities, reports a benefit-cost ratio of 9.0 for the $7.12 billion invested. The benefit-cost ratio rises to 17.3 when extrapolated through 2011 under the assumption that research benefits will continue to be realised at present rates. This confirms similar finding at INRA Morocco were every dollar invested in the development of Hessian fly bread wheat resistant cultivars, generates $9 additional benefit with profitability ratio in such investment of 37 per cent.

Plant breeding has been a central core of the research agenda at INRA and was the major driving force for funding research activities. The existence of a formal system of variety cataloguing and well organised private seed sector and adequate incentive policies for certified seed uses had a significant impact on the overall adoption of improved cultivars. It is not sure however that the allocation of resources counterbalanced such performance since there is no formal resource allocation par activity and the existing financial system provides very little flexibility to the overall management of INRA. Hence it always remains difficult to objectively assess breeding priorities in terms of predetermined resources allocation.

Nevertheless, in terms of human resources, research activities and outputs, we can easily assess that breeding of bread wheat, durum, barley for cereals, pate palm, olive and citrus, forage species such as oats and alfalfa and food legumes such as faba bean chickpea and lentils received the highest priority in the research agenda. Other species such as fruit trees (walnut, apricot, pistachios, pomes, pear etc.) and industrial crops such as sunflower, safflower, colza or sugar beet as well as vegetables occupied ambiguous position which largely depended on the research policy of the moment, the performance of researchers and the availability of finding resources.

CEREALS

Cereals are the most widely grown and consumed food crop in Morocco with an average of 5.3 million hectares which represents 60 per cent of total cultivated area. 90 per cent of cultivated area is rainfed. Barley is the main crop with 45 per cent of the total cultivated cereal area, bread wheat represents 30 per cent and durum 22 per cent (MADRPM, 2005). Cereal breeding has been the main research component of INRA breeding programmes. It is one of the strongest breeding programmes in the WANA region. A major breakthrough was achieved in the field of Hessian fly resistance. Genetic gain due to wheat improved varieties in rainfed area is about 1.55 per cent per annum and water use efficiency improved from 1.5 Kg/mm to 2.5 Kg/mm. Potential yields of released of bread wheat varieties was improved from 8 -12 quintals/ha to more than 75 quintals/ha. International centres such as CIMMYT and ICARDA played a key role in sustaining such achievements through provision of genetic resources and training of NARS scientists or technicians. Devolution of breeding activities to Morocco through decentralized selection of barley and wheat Hessian fly and rust resistance, helped foster further more the INRA role in the region. INRA scientists are actually posted at ICARDA and are playing a key role in the international wheat and barley breeding programmes.

Bread Wheat

Bread wheat breeding programme was initiated in 1921 at Rabat station. Early selection of genetic stocks was made among local landraces and introduced varieties from the Mediterranean region (Tunisia, Algeria) or other regions (India, Australia, USA, France and Italy). At that time barley (*Hordeum vulgare*) and durum (*Tricticum durum*) predominates the cropping system along with other species such as rye (*Secale cereal*), *Triticum monococcum*, *T. dicoccum* and *T. polonicum L.* (Jlibene, 2005). Then afterwards hybridisation was initiated to further exploring the existing genetic variability. Potential yield, rusticity and grain quality then starting from 1930, processing standards (W over150) used to be criteria of selection.

The first released cultivars were 4 varieties from Algerian origin, then 3 from Indian (Pusa type), 2 from Australia and 1 from France (Florence). Bread wheat cultivars such as "Cailloux" and "Florence" and "Florence Aurora" were among the most widely used cultivars during early century (INRA, 1948). During the 1960s, the first semi-dwarf genotypes and early maturing varieties of Mexican origin were introduced. The success of such varieties among which Siete Cerros and Potam, was limited and they were immediately discarded because of their susceptibility and the huge damage caused by *Septoria tritici.*

In 1973, the first high yielding cultivar "Nasma" which originated from a cross of Florence and Aurore C. was selected. This was a major breakthrough of the national breeding programme at INRA since it dominated the wheat production spectrum for more than two decades. Nasma is semi-dwarf genotype, it has a high potential yield and enhanced stability, it is resistant to lodging and to Leaf rust *(Puccinia recondita f.sp. tritici)* and has good grain quality (large seeded) that farmers appreciated most (Jlibene, 2005).

Since early 1980s, a wide breeding programme was initiated using CIMMYT varieties as parental lines in the hybridization programme. This resulted in release of more productive cultivar such as "Merchouch" and "Jouda" which in tern outweighed "Nasma". The CIMMYT genetic materiel used to have useful common features such as, earliness and good adaptation to Moroccan diverse conditions, probably resulting from the concept of mega-environments broad-based strategy of plant germplasm resources used at this center. Most materiel received from this center was spring or facultative wheat not sensitive to photoperiod and resistant to rusts. It has however two major problems, it is susceptible to septoria (Septoria *tritici, Mycosphaerella, graminicola*) and to Hessian fly (*Mayetiola destructor (Say)*). A specific breeding programme through hybridization was then initiated since 1987, one for semi-arid and the other for sub-humid conditions, with major objectives of improving resistance to Hessian fly and Leaf rust for the first environment and septoria and Leaf rust for the second. The USAID/MIAC project on dry land played a key role in fostering the breeding programme on Hessian fly resistance at INRA Morocco. That was a formidable challenge for agricultural science and for improved wheat production technologies of this century. Indeed Hessian fly does cause serious damage which is about 30 per cent loss and reaches 100 per cent in case of late sowing. The overall economical loss is about 200 millions $ per year. In 1997, two bread wheat cultivars "Aguilal' and "Arrehane" both resistant to Hessian fly were released to farmers.

Of the 32 genes expressing different level of tolerance or antibiosis against Hessian fly larvae, identified in the USA, ten genes (H5, H7H8, H11, H13, H14H15, H21, H22, H23, H25, and H26) were selected as conferring resistance to Hessian fly in the field in Morocco. Most of these genes are located on the A or the D genomes. The H22 gene has been successfully introgressed into adapted Moroccan bread wheat cultivars.

Alternative pest which are gaining importance are Sunn pest and Russian aphids. Accessions of *Triticum monococcum* from ICARDA combining both Hessain fly and Russian aphids resistance are being used in the breeding programme. Since 1995, a comprehensive breeding programme was initiated on drought tolerance and heat stresses using conventional breeding and most advanced tools in biotechnology.

Durum wheat

The objectives of the initial plant breeding programme were to improve productivity, response to high inputs and good pasta quality. Mountainous and oasis local land races provided basic genetic resources. Since 1930 five cultivars were released among which durum wheat "Zemarek", "Sbei" and "BD-2777" which continued to be grown in mountainous areas until recently because of its outstanding quality. These genotypes were however late maturing and susceptible to lodging.

During the 1960s, a new breeding programme was initiated through intraspecific hybridization with bread wheat and wild relatives for improving yield components, disease resistance to rust and Hessian fly. Such effort did not lead to any significant progress because of narrow genetic base for selection and depreciation of durum quality due to interspecific crosses. Since the 1970s, collaboration with CIMMYT then with ICARDA helped provide new genetic resources and led to increasing genetic variability of durum germplasm and led to release of two cultivars "Cocorit" and "Haj-Mouline". These varieties were high yielding and more adapted than previous cultivars but their quality limited their adoption in mountainous areas.

Since 1980, the breeding programme took a step forward through improvement of diseases resistance to root rote caused by *Fusarium sp*. and Helminthosporium (*Pyrenophora tritici-repentis*) and rusts: Stripe rust (*Puccinia striiformis*), Leaf rust (*Puccinia recondita*) Stem rust (*Puccinia graminis*). Selection within introduced germplasm and hybridization led to release eleven cultivars among which "Karim" from Tunisian origin, "Acsad-65" from "ACSAD", "Tassouat" and "Oum-rabia" from the national breeding programme.

As for bread wheat, Hessian fly became the most important common feature of the wheat breeding programme. Genetic resistance to Mediterranean Hessian fly biotypes has not been found initially in durum wheat, although large numbers of durum accessions were screened. Genes for resistance were found in common wheat, some of which were transferable to durum. Genetic base was further broadened through introgression of resistance genes from selected *Triticum araraticum* and *T. carthlicum* accessions using multiple backcross methodology.

This led to release of the most significant and the most significant result through registration of the first cultivars resistant to Hessian fly and tolerant to drought in 2003. Durum wheat grain is used to prepare various products including pasta and couscous. The acceptability of a durum variety is greatly influenced by its quality characteristics. These products generally require large vitreous kernels with high protein, good yellow pigment and strong to medium-strong gluten. Thus, the incorporation of quality characters essential for end-use products is now days the major challenge for durum breeding programme. In response to mil industry, a new project was initiated since 2005 in order to improve the yellow pigment related to the carotenoid and protein content of the endosperm.

Barley

About 80 per cent of barley grain production is used as animal feed on farms and the remaining is used as food. Morocco is ranked first worldwide for food and feeds barley consumption with the estimated average of 54.8 kg per capita consumed annually and is the 7th largest barley producer in the word. Morocco is considered one of the major centre of diversity for cultivated barley (*Hordeum vulgare* L.). This is evidenced by the large genetic variability found between and within barley landraces and existence of numerous wild relatives including populations of the wild progenitor of cultivated barley *H. vulgare ssp. Spontaneum* collected in the low Atlas region.

Barley breeding began in 1924 with the selection of lines from introduced two row barley accessions from the US, Australia and Europe as well as selection among sex row domestic landraces. The later programme led to selection and release of two cultivars "077" and "071" among local populations which are widely grown up today by most farmers. This clearly shows the importance of the genetic diversity in Morocco.

Several accessions have been introduced afterwards and a new breeding programme was initiated since 1970 for improving two-row barley which led to release of several cultivars among which Brasserie Maroc, Tamellalt and Asni. This type of genotype was not easily accepted by farmers who prefer the six-row barley though there is now evidence of their superiority. Since 1980, new accessions were introduced from the US in order to improved earliness and harvest index. Several accessions were also introduced from ICARDA with the main objective of improving forage and grain yields and disease resistance for leaf rust *Puccinia hordei,* powdery mildew *Erysiphe graminis f. sp.hordei*, net blotch *Pyrenophora teres f.sp*. teres and BYDV virus.

During this period there was a renewed interest for barley as a food component for human consumption. Its high soluble dietary fibre and â-glucan content compared with other cereals as well as the new prospect of hull-less barley genotypes that can be easily separated from the hulls after threshing could facilitate the use of barley grain for bread making and other nutritional foods. An additional programme to develop hull-less food barley varieties was initiated in 1985.Hull less barley helped initiated a wide food processing programme for diverse dietary uses. However, the difficulty of handling the seeds and their reduced seed-germination capability limited their inscription in the national catalogue.

Since 2003 a new breeding programme was initiated with two major components, one for marginal rainfed land (dry lands and high elevations areas), the second for favourable conditions. For the first production system, participatory breeding of local land races population, in situ-conservation and on farmer seed production are the main features of the breeding programme. For the second, the objective is to develop pure lines through conventional breeding methods within a long term strategy aiming at the introduction of male sterility for production of synthetic varieties.

Production of varieties with increased mineral and vitamins contents is another recent area of research lunched with collaboration with ICARDA. A new breeding project on biofortification aiming to mitigate the problem of micronutrients deficiency by improving zinc and iron contents, was recently initiated. But this programme remains relatively limited though its importance.

Maize

Cultivated area is about 267 000 ha 75 per cent is rainfed, 15 per cent irrigated and 10 per cent in mountains. Total rainfall in cultivated area is about 350 mm with 50 to 100 mm during the cropping season. Therefore drought is a serious limiting factor to corn production. Dew may however play a key role since most cultivated region area is in costal region. Initial breeding programme started at Rabat in 1947 with collection of 75 local land races. The bulk of this collection was constituted of early maturing populations with short stalks and ears, white or yellow grain, semi-flour or dent corn.

From 1947 to 1965 an extensive breeding programme was underway with introduction of nearly 256 lines and 281 varieties from various countries mainly US and France but also from various countries such as Portugal, Brazil and Spain. During this period the programme benefited also from an early FAO project aiming at introduction of US corn technologies in Europe. Breeding strategy evolved over that period as follows:

- Selection of corn varieties among local populations and introduced accessions;
- Selection of adapted hybrids among American corn hybrids;
- Creation of top- cross varieties resulting from hybridization of locale varieties (such as "Doukkala" and "Agouraï") and simple hybrids;
- Selection of double hybrids from simple hybrids;
- Creation of double hybrids with large genetic basis for rainfed agriculture.

High yield potential, adaptation to rainfed agriculture, earliness and dent yellow seeds were used as selection criteria. TX-21 was among the most important corn top-cross variety selected released during this period and continued to be used for almost two decades.

During the 1970s, the programme put more emphasis on top-cross varieties and simple hybrid production. Selection criteria included diseases resistance to rust (*Puccinia sorghi*) and corn-stalk borer, Sesamia (*Sesamia nonagrioides*).

Since 1980, a high priority was given to selection of variety populations for rainfed agriculture. Three varieties "Kamla", "Mabchoura" and "Doukkalia" in 1991 were released from such programme. A small project was also initiated in the dry land programme for production of synthetic populations but was resumed. Up to now 16 corn hybrids and 5 variety populations were registered in the catalogue. Yet their utilization remains problematic because of the overall situation of corn production. Some breeding work was also devoted to grain sorghum, but such work was neither systematic nor continued.

Triticale

Triticale varieties were initially introduced during the 1960s from CIMMYT. Adaptation trials carried out by INRA confirmed the superiority of triticales in costal sandy acid soils of favourable areas and in mountains (Mergoum and Kellida, 1997). During the 1980s cultivated area was about 10 000 to 15 000 ha. Actually the production of triticales is abandoned apart from few acres of forage mixture. The number of released cereal cultivars and the difficulty of commercialization have been two major handicaps for triticales production in Morocco. During the first period breeding objectives were high grain yield and adaptation to different agro-ecological conditions. Forage biomass and straw quality (Dry matter yield, haying requirement, regeneration after cutting and MAT) were afterwards included because of the increasing interest to triticales as forage crop. Screening for diseases resistance was carried out for rust, septoria and Hessian fly. Five cultivars of triticales were released in 1988 among which Juanillo and Beagle. A new dual purpose triticales, "Basma" for forage and grain production was released in 2002. Triticales still remains a secondary crop for cereal breeding though its high potential for chicken feed industry and forage production. An outreach regional breeding programme for Mediterranean areas was initiated with CIMMYT during the 1980s but this project was resumed afterwards.

Cereal Genetic Gain

In order to evaluate the genetic gain in cereal breeding we requested from a panel of INRA plant breeders currently working in Morocco or in International centers such as ICARDA to evaluate genetic towards achievement of breeding goals.

Table. Major breeding goals and the genetic progress*

Major crops	Major breeding goals	Genetic progress *
Bread wheat	♦ Potential and yield stability	1
	♦ Leaf rust resistance	2
	♦ septoria resistance	3
	♦ Hessian fly resistance	2
	♦ Russian aphids resistance	5
	♦ Multiple resistance	3
	♦ Drought tolerance	2
Durum wheat	♦ Potential and yield stability	2
	♦ root rote and Helminthosporium	3
	resistance	2
	♦ rusts resistance	1
	♦ Hessian fly resistance	3
	♦ Drought tolerance	4
	♦ Quality (yellow pigments and protein)	3
	♦ Multiple resistance	3
	♦ Specific adaptability (High lands)	
Barley	♦ Potential and yield stability	2
	♦ leaf rust resistance	2
	♦ powdery mildew resistance	3
	♦ net blotch resistance	3
	♦ BYDV virus resistance.	4
	♦ Feeding values	2
	♦ Biofortification (Zn, Fe)	5

FOOD LEGUMES

Food legumes are important pulse crops in the Moroccan agriculture. Total cultivated areas is about 350 000 hectares, 43 per cent of which is Faba bean, 19 per cent chickpea, 14 per cent lentil and 9 per cent peas. They are important source of protein and they play an important role in the cereals rotation system. This role becomes however limited because of the decline of cultivated areas and the overall changes in human diet.

Faba Beans

The major objective of early faba bean breeding programme in 1949 was to collect and identify genetic source of high yielding varieties with acceptable seed size among local land races and introduced accessions from Spain or Italy. The result was the release during the 1950s of six cultivars of faba bean minor and major and collection of more than 200 accessions.

In 1979, the food legume breeding programme knew fundamental changes through collaboration with ICARDA. In 1988, the CGIAR decided to develop an outreach faba bean programme in Morocco and transferred the ICARDA faba bean breeding programme to INRA. This programme was resumed in 1991 and collaborative projects were developed afterwards with GTZ for the North Africa region, through networking projects such as REMAFEVE, REMALA and Parasitic weeds.

Breeding for high yield and tolerance to orobanche (*Orobanche crenata* Forsk), resistance to botrytis (*botrytis faba*) and nematodes (*Ditylenchus dipsaci*), became the most challenging issues to the faba bean breeding. Orobanche crenata, a parasitic weed, has been the most limiting factor for faba bean production. In Morocco, infestation of Orobanche in cultivated food legumes crops, evolved from 12 per cent in 1981, 26 per cent in 1994, to 30 per cent in 2001 and reached 51 per cent in 2003 with estimated average yield losses in 1998 of 37.4 per cent.

However, the expected progress remains very slow because of the major difficulty for combining quantitative characteristics involving resistance to Orobanche, yield per plant and large seed size (Cubero and Hernandez, 1991). Today, in spite of some genetic progress (release of one line Sel.88 Lat 18105, faba bean minor tolerant to orabanche), the main challenge remains however is the introgression of multiple resistance to both orobanche and botrytis in large seeded type.

Chickpea

Chickpea is the second most important legume. It is cultivated as spring crop. Early breeding programme, initiated since 1920, put more emphasis on germplasm collection of spring chickpea of large seeded Kabuli types. Breeding work was strengthened in 1943 with germplasm collection of both kabuki and specific Desi-types maintained during winter season for ascochyta blight evaluation. Do the plant breeder at that time was aware of the potential of winter chickpea? There were no evidence for this, nevertheless a winter hardy types were maintained at this collection and two spring chickpea of Kabuli type, "PCH34" and "PCH 37", were released from such programme. The requirement of larger seed size and competition of Mexican and Spanish types in the commerce, were a serious handicap for any further progress in the breeding programme.

In 1978, INRA initiated a new breeding programme on horizontal resistance of spring chickpea to ascochyta blight (*Ascochyta rabiei (Pass.)*) with FAO collaboration. Extensive works on screening and breeding methodologies were developed within the frame work of such programme. About 26 lines with durable resistance were selected from this programme. However, the specific nature of blight pathogenic variability in Morocco, limited the release of such breeding material. It is worth mentioning that such programme was afterwards extended to cover durable resistance of Faba bean to orobanche and tomato to Fusarium wilt, but this programme was not continued.

The renewed interest for chickpea became however very clear through collaboration with ICARDA-ICRISAT programme posted at ICARDA since 1979. The introduction of the concept of winter chickpea completely changed the configuration of the breeding programme. Winter planting in November of winter hardy genotypes was suggested as an alternative option for increasing yield and stability in semi-arid environments. Resistance to Ascochyta Blight become a prerequisite in such situation because of the major blast of the disease during the winter season. Potential yields of winter chickpea was increased to more than 2.0 t/ha as compared with 0.6 t/ha of spring chickpea. This represents an overall increase of nearly 210 per cent with an earliness of 25 to 45 days.

Two varieties ILC 482 and ILC 195 were released in 1987 but their seed size was too small to be accepted by farmers. Further collaboration with ICARDA led to release since 1992 of 5 new winter chickpea cultivars with larger seed size among witch "Douyet" and "Moubarak". The high seed cost of such cultivars has been one of the major handicap for any further uses of this genetic materiel by farmers.

In view of the lesson learned farmers' field and difficulty to recognize winter sown chickpea from spring type and the high risk for planting the latter at early season, a new breeding approach was initiated in 1987 aiming at selection of dual season chickpea adapted for both spring and winter seasons.

Lentil

Breeding programme of this crop followed the same pattern, with release of two cultivars L26 and L56 large seeded type from the early works. Intensive breeding programme was initiated with ICARDA, then with Washington State

University-USDA. The main objectives of the new breeding programme were to improve the yields, resistance to rust (*Uromyces viciae-fabae)* and earliness. The first resistant cultivars to rust "Bakria" named after "Precoz" from Argentina origin, was released in 1989 and the first a cross bred cultivar "Chaouia" of Laird and Precoz, was released in 2004. The overall increase of yield potential was above 130 per cent and earliness by three weeks in addition to rust resistance. The later character was the major accomplishment during the last twenty years and significant yield improvement was achieved through selection on new inbred lines.Unless other food legume crops, there has been a wide spread of the new released cultivars in most cases through informal channels from farmers to farmers seed production. Massive importation of lentils from Canada is however causing serious traits to local lentil production and will seriously affect the overall impact of INRA achievements.

Food Legumes Genetic Gain

To conclude, major achievement towards breeding goals in food legumes could be summarized as follows.

Table. Major breeding goals ad genetic progress*

Major crops	Major breeding goals	Genetic progress *
Faba beans	♦ Potential and yield stability	2
	♦ Botrytis resistance	4
	♦ Rusts resistance	3
	♦ Nematode resistance	5
	♦ Orobanche resistance	3
	♦ Multiple resistance (botrytis&orobanche)	5
Lentil		
	♦ Potential &yield stability	2
	♦ Rust resistance	2
	♦ Earliness	2
	♦ Mechanical harvesting	3
	♦ Cooking quality	3
Chickpea		
	♦ Potential &yield stability	1
	♦ Ascochyta blight resistance	2
	♦ Fusarium wilt resistance	4
	♦ Seed size	3
	♦ Winter adaptation	2
	♦ Dual season	3

FORAGES

Feeding sources in Morocco are: crop residues and fallow (35.9 per cent), natural pastures (21 per cent), cereals mainly barley (15.7 per cent), forage fodder (14.2 per cent) and industrial by-products (13 per cent). Rangelands and unproductive lands are about 52,900,000 ha. Fodder forage are grown on approximately 386 000 hectares, this includes annuals such as oats (19.5 per cent), fodder barley (26.5 per cent), berseem (12.6 per cent), and fodder maize (4.5 per cent), and perennial Lucerne is about 22.8 per cent (Chaouki et al, 2005).

Forage and pasture seed production is less developed compared to other crops. This is due probably to the fact that forage seeds are considered by farmers as by-products. Therefore, most seeds are produced and distributed through the informal channels. The most important forage crops are Lucerne, berseem clover, barley and maize under irrigation; oat, vetch, barley, ryegrass, fodder peas, and annual medic in the rainfed zones. Perennial grasses (*Dactylis, Festuca, Agropyron*, etc.), annual medics, clovers and fodder shrubs (*Atriplex, Acacia, and native shrubs*) are important alternatives for range improvement.

Primary works on forage breeding during colonial period, began with extensive collection and characterization of pasture species and fodder crops for the cattle Production Systems. Important documentations were reported on species such as Trifolium, Lucerne, medicago, crotalaria and lupine, oats and other species. But it was until 1980 that the forage breeding programme knew the most important changes in the overall research strategy based on commodity programme within the frame work of a long term cooperation with GTZ. During a first phase of such cooperation, research put more emphasis on pasture forage for mixed livestock small ruminant livestock production systems. Breeding works concerned the following categories of forage species (Alfaïz, 2003):

- Pastoral leguminous species (Medicago spp., Trifolium subterraneum, Trifolium fragiferum, Lotus corniculatus);
- Pastoral perennial grasses (Ehrharta calycina, Festuca arundinacea, Phalaris aquatica, Lolium rigidum);
- Fodder crops such as oats, maize, Lucerne and beta forage;
- Forage mixture species cultivated in associations such as vetch, peas, oats and triticales.

Breeding strategies covered several aspects depending on the species:

- Selection among imported pre-selected material from international Research Centres or from bilateral collaborations;
- Selection in segregating material after target crosses;
- Selection of local collected ecotypes;
- and introduction of selected varieties from other countries.

Major selection criteria include forage biomass and grain yields, forage quality, palatability, anti-nutritional factors and diseases resistance. During this phase significant progress were achieved through selection of local biotypes of Medicago spp and Trifulium subterraneum adapted to acid soils better than Australian ecotypes used in lay farming operation. Along with the regular breeding programme, a major breakthrough was achieved through collection and evaluation of local forage genetic resources with collaboration with GTZ, ICARDA and Centre for Legumes in Mediterranean Agriculture (CLIMA) at the University of WA (UWA) in Australia. Collected accessions from 1993 to 2005 are summarized in table. The GTZ project helped install the first long term storage facilities of INRA in Rabat centre. During the second phase the breeding programme put more emphasis on traditional forage cultivated species readily acceptable by farmers. Priority was then allocated to Oats, Lucerne, and lupine, peas, vetch and forage mixture.

Oats

Oats (*A. sativa*) used to be a traditional crop for the breeding programme though it gained importance during the last fifteen years. Oats were introduced to Morocco by the French at the beginning of the twentieth century. At that time, most forage came mainly from natural grazing, stubbles, fallow, Lucerne and barley. There were several reasons for the success of oats: it was a crop that French farmers knew well, and yielded several products, such as grain and forage, or could be mixed with vetch. It was also in particular immune to the Hessian fly that caused serious damage to other cereals (Grillot, 1939). In Morocco, Oat is grown essentially in rainfed areas, almost exclusively for forage hay, alone or as a mixture with annual legumes such as vetch. The current areas are about 70 000 ha for oats and 50 000 ha for oat and vetch mixture. In recent years, there has been an increase interest for oats pure cropping resulting from the difficulties for practicing forage mixtures, non-availability of adapted and synchronized varieties, weed control, high cost and scarcity of vetch seed.

The first varieties from Europe did poorly in Morocco. Thereafter, Grillot (1939) introduced "Byzantina" types from Algeria, which constituted the genetic basis for the first locally selected varieties. Some of these varieties are still grown, such as "Roummani", "Zhiliga" and "Tiddes", but they are now very susceptible to the common diseases and cannot cover all the potential ecological zones for oats. Locally produced oat seeds are mainly a mixture of these varieties, with some imported admixtures. The actual breeding oats programme aims to select cultivars adapted to three ecological areas, favourable rainfed, semi-arid and highlands. The following criteria for selection have been used:

From this breeding programme, seventeen cultivars were registered in the national catalogue and released to farmers. Among such cultivars "Rommani" realised in 1982, "Ghali", "Faras" and "Soualem" in 1989, selected from introduced accessions and "Bounejmat" and "Allal" released in 2003, from national hybridization programme. Significant progress has been mad in breeding for diseases resistance such as rusts and powdery mildew, yet resistance to BYDV remains one of the biggest challenge to oat breeding. Morocco is a centre of genetic diversity of the genus Avena. According to the morphological species concept, 32 species have been named and more recently 13 species recently identified using criteria for biological species. During the last ten years, an important breeding programme was initiated in order to make use of the full potentials of wild tetraploids oats, A. *murphyi* and *A. magna*, as resources of protein and disease resistance to rust, powdery mildew and BYDV.

Forage oats for biomass production	Grain oats
- Dry mater yield; - Forage quality: digestibility,% MAT, palatability, silage and Haying properties; - Tolerance to the most important diseases of oats in Morocco: stem rusts (Puccinia graminis Pers. f.sp. avenae), and Barley Yellow Dwarf Virus (BYDV); - Resistance to lodging and shattering; - Earliness for semi-arid areas, mid-earliness to late for favourable rainfed areas, and late for the mountains	- Grain yields; - Quality: protein content, metabolic energy, beta-glucan content; - Resistance to rust, to BYDV and to Crown rust, caused by Puccinia coronata f.sp. avenae

Eight hexaploid populations resulting from a cross of (*A. sativa x A. magna*) *x A. sativa*, have been selected and the first cultivar resulting from this cross may be released soon (INRA, 2005).Such breeding programme constitutes one of the major breakthroughs in the forage breeding programme at INRA. Release of high protein oat cultivars will not only be of great importance to the Moroccan feed industry, but also to the world oat markets. Despite such effort, improved certified seed is only about 10 to 20 percent of oat cultivated area. This is mainly due to competition from the common seed trade, which seems to be more profitable.

INDUSTRIAL CROPS

Fibre Crops

Industrial crops include textile fibre crops and oil seed crops. The first category received a highest priority during the colonial and post colonial periods which recognised the importance of fibre crops as alternative crops for export markets and for responding to local needs. Traditional culture of cotton and cannabis were reported in Marrakech and Sefrou Area as well as Ramie in Souss and Gharb regions. Fibre crops are field crops grown for their fibres, which are used to make paper, cloth, or rope.Cotton (*Gossypium hirsutum* L.) breeding programme was established in 1920 in Rabat then after ward at Tadla stations. Selected lines included both short and long or extra-long-staple. Selection criteria included evaluation of cotton tolerance to Verticillium wilt (*Verticillium dahliae* Kleb.) and avoidance of caterpillars (*Helicoverpa armigera)* through selection of early maturing varieties.

Early breeding work put emphasis on germplasm introduction from various countries including Niger, USA, Egypt, Sudan and Latin American countries.

Earliness, yield, boll size, seed index, lint percentage, fibre length, fibre strength, and micronaire were the main features of cotton selection. Selected cultivars in 1951 included Karnak, Malaki and Pima cotton 67 an American-Egyptian variety which was widely used in cotton cultivation in Morocco for more than four decades. Hybridization programme was also initiated using Pima 67 as a parental line (Bourg, 1951). New released cultivars in 1989 included eight cultivars of "Tadla type" most of them from Egyptian origin. The programme was however closed during the 1990s because of the decline of cotton production in Morocco. Linseed or Flax (*Linum. usitatissimum L.*) was also an important textile crop in early breeding works during the same phase. Traditional cultivation of local land races of oil linseed was mentioned in Chaouia region in Morocco. Selection was carried out both for linseed oil and flax fibre or mixed types. Selection criteria included fibre yield and quality, resistance to rust (*Melampsora Lini*) and self-pollination for breeding purposes. During the 1940s and 1950s, genetic improvement concerned selection of local populations (exp. Line 0196), introduction of new accessions from USA, Holland, Italy and Russia (exp. varieties "Costrama" and "Italie") and hybridisation as well as mutation breeding (Bourg, 1951).

Early breeding work included also collection of other species such as hemp (*Cannabis indica*), ramie (*Boehmeria nivea*) and secondary textiles such as jute (*Corchorus olitorius* and *Corchorus capsularis*) and hibiscus (*hibiscus cannabinus L.*). This programme was discontinued during the 1970s because of difficulty of the fibre extraction technology and the change in the breeding priorities which put emphasis exclusively on cotton improvement. Most of this variable genetic materiel was lost because of lack of adequate long term storage facilities during that period.

Oilseed Crops

Oilseed crops such as sunflower, safflower and rapeseeds were the second most important species of industrial plants. Breeding work on sunflower (*Helianthus annuus L.)* was initiated since 1958 as the major oilseed crop in Morocco with the introduction of two lines Vniimk and Stepniak of Russian origin. During the period of 1960s and 1970s, these two lines along with a new line called "Peredovik" have been used in the initial genetic improvement programme with a German cooperation. Since then, two cultivars with high oilseed yields per hectare, "Oro-9" (renamed karima) and "Record" (named afterwards Salima) were released (Nabloussi et al, 2005). Till now, these two cultivars dominate sunflower cultivation in Morocco.

During the 1980s and early 1990s, an important breeding programme on sunflower breeding was developed within the frame work of a World Bank project which included several commodities among which oilseed crops. Two major breeding projects were established one for selection of multiple-line cultivars devoted to rainfed conditions and production of hybrids resistant to downy mildew (*Plasmopara halstedii* Farl.) for high input agriculture. A new approach using the same principle of winter chickpea cropping, was also initiated though selection for cold tolerance for early cultivation of sunflower. In 1997 a new winter type "Ichraq" and a hybrid "Manar" resistant to mildew were released. The sunflower breeding programme was resumed since 2003 because of the decline of oilseed production in Morocco.

Some selection work was also carried out on other oilseed crops such as Rapeseed (*Brassica napus*) and safflower (*Carthamus tinctorius)* during the same period through introduction of germplasm, but this programme was discontinued although these species are gaining again some importance. A specific contract was lately signed with the ministry for agriculture in 2006, for further improvement of safflower spineless varieties.

HORTICULTURE CROPS

Date Palm

Date palm (*Phoenix dactylifera* L.), plays key social, economical and ecological roles in the Oasis production systems. It helps prevent desertification, provide general income to poor farmers and unsure protection for sub-adjacent crops such as trees, cereals and Lucerne. The total cultivated area is about 48 000 ha with 4.8 millions palm trees covering 90 oasis. Genetic profile of date palm groves is composed of local population mixtures "Khalt" which represents 47.5 per cent of total cultivated area and other local cultivars with good quality such as "Boufeggous" (12 per cent), "Jihel" (12 per cent), "Bousekri" (2 per cent) and the noblest variety "Mejhoul" (0.3 per cent).

These genetic resources resulted from natural selection and endogenous knowledge of farmers which contributed mostly to the genetic advance of date palms in Morocco and in the region. Most of these cultivars are however susceptible to Beiyoud disease, caused by a soil born fungus *Fusarium oxysporum* susp. *Albedini.* This fungus represents the main threat to date palm production since it has killed more than 12 millions trees and continually causing damage to 4.5 to 12 per cent of date palms (Djerbi et al, 1986). Date palm breeding proved to be fastidious and time consuming since it takes almost 30 years to make three back-crosses to obtain first offshoots from crosses. The dioecism nature of date palm is another obstacle, since it would require almost eight years for the female seedlings to bear fruits.

As we will say however, the overall breeding strategy followed at INRA Morocco during the 1980s proved to be efficient for speeding up the process of selection. Modern biotechnology provided new opportunities for further genetic improvement. The FAO regional research project started during the period of 1980-85 on date palm for North Africa and Middle East, played a key role in fostering further more the overall research agenda on date palm in Morocco and helped develop the *in-vitro* propagation techniques and modern screening methods for beiyoud resistance. It also helped provide base line studies to the actual research programme on molecular markers for beiyoud characterization and disease resistance.

Historically, date palm breeding started during the period between 1949-56 with selection of resistant clones to Beiyoud disease from local populations and initiation of hybridization programme. Unfortunately most of the breeding documents were lost following interruption of such programme which was resumed afterwards through collaboration with CIRAD and INRA France in 1963. Since then, date palm breeding programme continued staidly with the main objectives of improving disease resistance to beiyoud and quality. The overall breeding strategy changes could be summarized as follows:

- Clone and variety selection for resistance to beiyoud from local date palm groves. For this purpose, the first systematic collection among "Khalt" populations was initiated between 1963 and 1969 and continued afterwards during the 1980s. Cultivars with acceptable level of resistance selected from Draa valley included: "Bou Sthammi noir", "Bou Sthammi blanc", "Tadment", "Iklane", "Sair Laylet", "Bou Feggous" and "Moussa" The fruit quality of these varieties is very poor and limited their extension. However, recent research on technology processing, showed that the commercial value of some of these varieties may be enhanced through industrial processing for marmalade or syrup production;
- Individual selection among female plants combining both resistance to beiyoud and good quality from endogenous populations' mixtures "Khalt" also called "Saïrs". About 2 337 "Saïrs" with good quality were collected from 1967 to 1986. Promising materiel combining both quality features and resistance to beiyoud have been selected from this genetic pools;
- Hybridization programme was initiated in 1972 and continued during the 1980s through introgression of resistance genes in cultivars with good quality such as "Mejhoul" "Bou Feggous" and use of backcross-recurrent selection for improving furthermore the overall quality. 15 clones combing both resistance and good quality features proved to be promising from this programme;
- Use of modern biotechnology approach resulting from extensive research during the 1980s and 1990s on tissue culture, *in vitro* selection and identification of molecular markers both for beiyoud resistance and characterisation of pathotype variability of the fungus.

From this overall breeding programme, new date palm clones resistant to beiyoud disease and good quality were released since 2000. Actually about 200

000 *vitroplants* of cultivars "Najda", "Al-Amal", "Bourhane" and "Al-fayda" were delivered to farmers. Such result represents the second most important world wide achievement for date palm after the Hessian fly resistance breakthrough in cereals.

Olive

The olive (Olea europaea L.) is well known in the Mediterranean area. It yields two products, table olives and olive oil, both of which are important commodities on world markets. Total cultivated area is about 500 000 ha with 50 millions trees which put olive up front as the most strategic horticulture crop in Morocco. The national olive plan gave a high priority to this crop and it is expected that the total area will be almost doubled by 2020. Variety profile is dominated by a local population "Picholine marocaine" which represents nearly 98 per cent of the total cultivated area. Recent study showed that this assumed population was rather a single major cultivar with substantial genetic variability probably resulting from much local domestication Khadaril et al, 2007). Picholine marocaine has however two major problems, its alternate production and susceptibility to fungus Spilocea *oleaginum*. Early work on olive breeding started in 1927 through introduction from various Mediterranean countries, but mostly from Italy and Spain, of 120 varieties at Menara-Marrackech and Meknes-Ain Taoujdat experimental stations. The first selected cultivars which are used up to now but in limited area were: "Frantoio" and "Ascolana Dura" from Italy, "Gordale" and "Manzanilla" from Spain and "Picholine Languedoc" from France. The first cultivar is oil variety the others are double purpose or table olives.

During a second phase which started in 1978 up to 1994, olive breeding put more emphasis on clone selection among endogenous local olive populations. Early collections started from the Northern region, followed by the central region (Haouz and Tadla), then in more recent years in the southern regions. The major objectives were selection of improved olive clones through exploration of the wide genetic variability of "picholine marocaine". Recent studies showed individual variability within this cultivar of 22.3 per cent up to 56.4 per cent for yield performance depending on localities.

From this overall programme, "Haouzia" and "Ménara" clones were released during the 1990s followed by clones named M26, K26 and S19. Early entrance to production starting from the third year, high oil content with an average increased productivity of 100 per cent over "picholine marocaine", are the main characteristics of this genetic material. About 8 millions plants of "Haouzia" and "Ménara" are actually planted in Morocco. These clones are also planted in other countries such as Spain, Australia and United Emirates. During the 1990s, a new phase was lunched through genetic hybridization, with the objective of improving oil content, potential yield and resistance to major diseases. Such breeding programme was initiated within the frame work of a regional breeding project with the international olive centre (COI). About 1 890 progenies resulted from 19 crosses among two gene pools composed of "Picholine marocaine", "Ménara", "Haouzia", "M26" on one hand and "Arbequine", "Manzanille", "Picholine du Languedoc", "Leccino" on the other. Selected lines from these crosses included two high yielding and oil content genotypes (40 to 50 per cent DM) with higher oleic acid content, four cultivars for table olive and five double propose olive with almost 100 per cent performance over the picholine. This result was made possible through reduction of the juvenile phase which is one of the major handicaps for olive breeding.

It is worth mentioning that since 2002, INRA hosted with the help of COI, an international olive collection at Tassoaut station for collaboration with Mediterranean countries.

Citrus Crops

Morocco is one of the most important citrus producing countries in the Mediterranean region. Historical arguments show that citrus fruits have been known for many centuries and some believe they were introduced into the country by the Romans. For centuries, however, citrus was grown in the country as scattered trees for local consumption. Some local citrus ecotypes are very famous in Morocco among which, mandarins of Tanger, lemon "Limoun Boussara" and "Cedrats" of Marrakech.

Moroccan citrus growing area is of 77 400 hectares with a total production of 1.2 to 1.5 million tons depending on the season. This represents nearly 1.3 per cent of total word citrus production of which 42 per cent is exported mainly to Western Europe. Current citrus production planning puts more emphasis on small fruit citrus crops. About 20 genotypes are used in farmers citrus groves, among which five dominate the overall sector. These include orange type "Maroc-late" (35 per cent) and "Navel" (21 per cent) as well as "Clémentine" (27 per cent) and. The rest is composed of mid-season cultivars of blood orange (Sultana and Sanguine) and new small fruit clones of mandarins Ortanique "Nour", "Afourer" and "Nova".

Sour orange, *Citrus aurantium* L., is practically the only rootstock used in the country. Some Troyer citrange was used but it does not produce a high performance tree in calcareous soils and is being abandoned. Therefore diversification and selection of alternative rootstocks remains one of the major breeding objectives for breeding improvement.

Cultivars introduced from different countries or locally selected resulted in a large genetic variability of citrus in Morocco. The first systematic review of existing and potential genetic resources was described in 1968. A Collection was established since 1930s in "Souihla" research station, near Marrakech, with actually more than 600 accessions and the second at El Manzah experimental station near Kenitra since 1964 with a collection of more than 400 accessions maintained actually, followed by Afourer station collection in 1964 with 80 accessions and Allal Tazi collection in 1972 with 100 accessions.

Breeding programme covered the following aspects:

- Selection of nucellar clones of commercial citrus varieties and releasing healthy budwood to citrus nurseries for commercial propagation;
- Virus free germplasm banks;
- Genetic improvement of main citrus varieties and rootstocks
- Characterization of major citrus diseases such as Trisetza, Psorosis, Exocortis viruses, stubborn disease and Phytophtora fungus (Phytophtora gummosis)
- Indexing procedures as well as shoot-tip grafting protocols for productions of virus-free genetic material were also developed.

An important hybridisation programme was also initiated during the 1980s with a more systematic characterisation and documentation of genetic resources. Selection priorities for the varieties have been since, oriented towards a longer harvest period (early maturing or late maturing cultivars), displaying higher pomologic and organoleptic qualities (easy-peel seedless mandarins for example), disease resistance (tristeza, greening, *Phaemularia angolensis*) and high yield. Other criteria of selection include for juice processing quality features such as high sugar content.

Rootstocks selection priorities include tolerance and/or resistance to various strains of *tristeza* virus and *Phytophtora* fungus, tolerance to high salinity and calcareous soils. In this regard a project was developed in order to enhance the efficiency of citrus germplasm utilization by rootstock breeding through sexual recombination and somatic hybridization, within the frame work of a Euro-Mediterranean initiative on citrus rootstock with collaboration of CIRAD (France). Since 2000, an important programme using advanced biotechnology tools was initiated. Apart from production of free virus plants for citrus production nurseries of common commercialized varieties and rootstocks, selection of mandarin cultivar "Nadorcott" has been the major breakthrough of INRA citrus programme. Nadorcott mandarin was selected from a population of mandarins planted in 1964 at INRA experimental station near the village of Afourer, Morocco. It was selected due to its lateness (production up to April), ease of peeling and reddish rind colour when compared with Murcott and it was later discovered that seedless fruit were produced when no pollination occurred. Initially, the selection "INRA-W21" was called Afourer, which is now used as a trade name for Moroccan produced fruit. Subsequently, it was named Nadorcott. This variety The Nadorcott variety is protected with a patent since October 2004. This variety is grown in several countries including South Africa, Spain and USA.

10

Chromosomal Techniques and Evolution in Plant and Trees

CHROMOSOMES BEHAVIOUR DURING MEIOSIS AND PARTIAL LINKAGE

Partial linkage was discovered in the early 20th century. The cross shown here was carried out by Bateson, Saunders and Punnett in 1905 with sweet peas. The parental cross gives the typical dihybrid result, with all the F_1 plants displaying the same phenotype, indicating that the dominant alleles are purple flowers and long pollen grains. The F_1 cross gives unexpected results as the progeny show neither a 9: 3: 3: 1 ratio (expected for genes on different chromosomes) nor a 3: 1 ratio (expected if the genes are completely linked). An unusual ratio is typical of partial linkage. The critical breakthrough was achieved by Thomas Hunt Morgan, who made the conceptual leap between partial linkage and the behaviour of chromosomes when the nucleus of a cell divides.

Cytologists in the late 19th century had distinguished two types of nuclear division: mitosis and meiosis. Mitosis is more common, being the process by which the diploid nucleus of a somatic cell divides to produce two daughter nuclei, both of which are diploid. Approximately 10^{17} mitoses are needed to produce all the cells required during a human lifetime. Before mitosis begins, each chromosome in the nucleus is replicated, but the resulting daughter chromosomes do not immediately, break away from one another. To begin with they remain attached at their centromeres and by cohesin proteins which act as 'molecular glue' holding together the arms of the replicated chromosomes.

The daughters do not separate until later in mitosis when the chromosomes are distributed between the two new nuclei. Obviously it is important that each of the new nuclei receives a complete set of chromosomes, and most of the intricacies of mitosis appear to be devoted to achieving this end. Mitosis illustrates the basic events occurring during nuclear division but is not directly relevant to genetic mapping. Instead, it is the distinctive features of meiosis that interest us. Meiosis occurs only in reproductive cells, and results in a diploid cell giving rise to four haploid gametes, each of which can subsequently fuse with a gamete of the opposite sex during sexual reproduction. The fact that meiosis results in four haploid cells whereas mitosis gives rise to two diploid cells is easy to explain: meiosis involves two nuclear divisions, one after the other, whereas mitosis is just a single nuclear division.

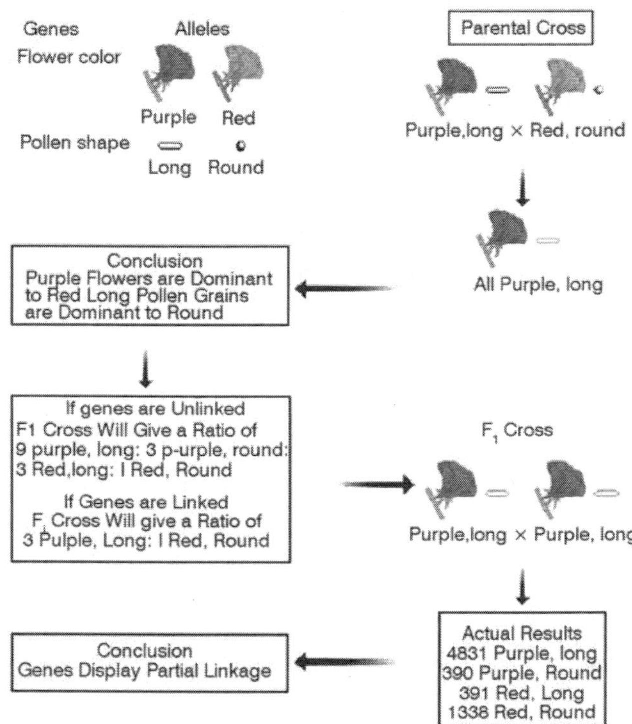

Fig. Partial Linkage.

This is an important distinction, but the critical difference between mitosis and meiosis is more subtle. Recall that in a diploid cell there are two separate copies of each chromosome. We refer to these as pairs of homologous chromosomes. During mitosis, homologous chromosomes remain separate from one another, each member of the pair replicating and being passed to a daughter nucleus independently of its homolog. In meiosis, however, the pairs of homologous chromosomes are by no means independent.

During meiosis I, each chromosome lines up with its homolog to form a bivalent. This occurs after each chromosome has replicated, but before the replicated structures split, so the bivalent in fact contains four chromosome copies, each of which is destined to find its way into one of the four gametes that will be produced at the end of the meiosis. Within the bivalent, the chromosome arms (the chromatids) can undergo physical breakage and exchange of segments of DNA. The process is called crossing-over or recombination and was discovered by the Belgian cytologist Janssens in 1909. This was just 2 years before Morgan started to think about partial linkage.

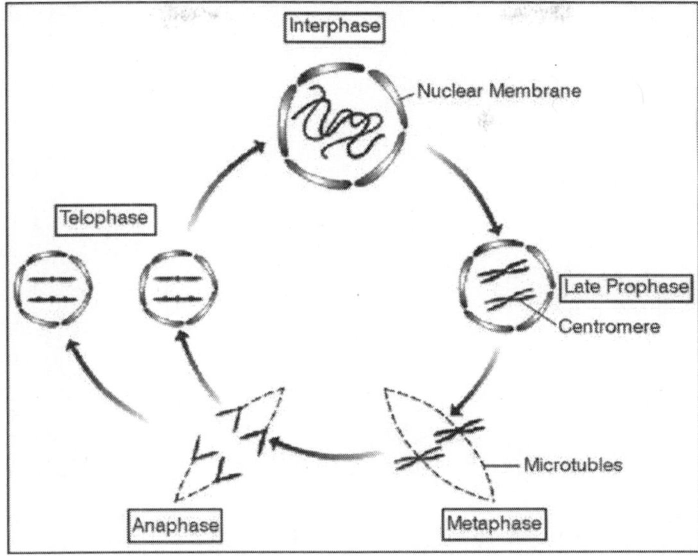

Fig. Mitosis. During Interphase the Chromosomes are in their Extended Form.

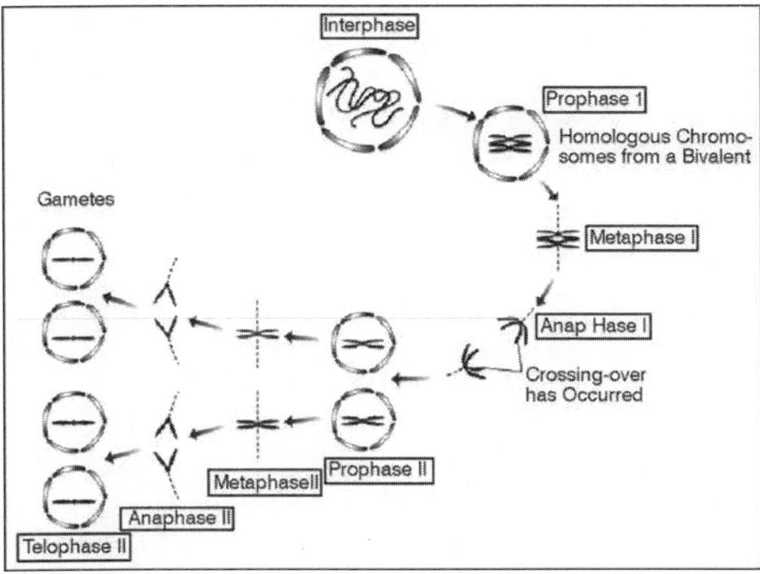

Fig. Meiosis. The Events Involving One Pair of Homologous Chromosomes are shown; At the Start of Meiosis the Chromosomes Condense and Each Homologous Pair Lines up to form a Bivalent.

How did the discovery of crossing-over help Morgan explain partial linkage? To understand this we need to think about the effect that crossing-over can have on the inheritance of genes. Let us consider two genes, each of which has two alleles. We will call the first gene A and its alleles *A* and *a*, and the second gene B with alleles *B* and *b*. Imagine that the two genes are located on chromosome number 2 of *Drosophila melanogaster*, the species of fruit fly studied by Morgan.

We are going to follow the meiosis of a diploid nucleus in which one copy of chromosome 2 has alleles *A* and *B*, and the second has *a* and *b*. This situation is illustrated in *Figure*. Consider the two alternative scenarios: The following drawing Shows a Pair of Homologous Chromosomes, A and B are Linked Genes with Alleles *A*, *a*, *B* and *b*. On the left is a Meiosis with no Crossover between A and B: two of the Resulting Gametes have the Genotype *AB* and the other two are *ab*. On the right, a Crossover Occurs between A and B: the four Gametes Display all of the Possible Genotypes: *AB*, *aB*, *Ab* and *ab*

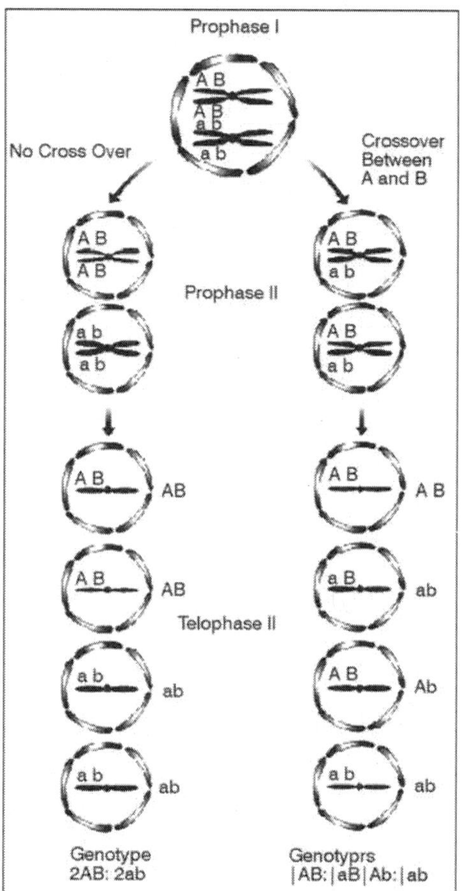

Fig. The Effect of a Crossover on Linked Genes.

1. *A crossover does not occur between genes A and B.* If this is what happens then two of the resulting gametes will contain chromosome copies with alleles *A* and *B*, and the other two will contain *a* and *b*. In other words, two of the gametes have the genotype *AB* and two have the genotype *ab*.
2. *A crossover does occur between genes A and B.* This leads to segments of DNA containing gene B being exchanged between homologous chromosomes. The eventual result is that each gamete has a different genotype: 1 *AB*, 1 *aB*, 1 *Ab*, 1 *ab*.

Now think about what would happen if we looked at the results of meiosis in a hundred identical cells.

If crossovers never occur then the resulting gametes will have the following genotypes:

- 200 AB
- 200 ab

This is complete linkage: genes A and B behave as a single unit during meiosis. But if (as is more likely) crossovers occur between A and B in some of the nuclei, then the allele pairs will not be inherited as single units. Let us say that crossovers occur during 40 of the 100 meioses.

The following gametes will result:

- 160 AB
- 160 ab
- 40 AB
- 40 aB

The linkage is not complete, it is only partial. As well as the two parental genotypes (*AB*, *ab*) we see gametes with recombinant genotypes (*Ab*, *aB*).

Partial Linkage to Genetic Mapping

Once Morgan had understood how partial linkage could be explained by crossing-over during meiosis he was able to devise a way of mapping the relative positions of genes on a chromosome. In fact the most important work was done not by Morgan himself, but by an undergraduate in his laboratory, Arthur Sturtevant. Sturtevant assumed that crossing-over was a random event, there being an equal chance of it occurring at any position along a pair of lined-up chromatids. If this assumption is correct then two genes that are close together will be separated by crossovers less frequently than two genes that are more distant from one another. Furthermore, the frequency with which the genes are unlinked by crossovers will be directly proportional to how far apart they are on their chromosome. The recombination frequency is therefore a measure of the distance between two genes. If you work out the recombination frequencies for different pairs of genes, you can construct a map of their relative positions on the chromosome.

It turns out that Sturtevant's assumption about the randomness of crossovers was not entirely justified. Comparisons between genetic maps and the actual positions of genes on DNA molecules, as revealed by physical mapping and DNA sequencing, have shown that some regions of chromosomes, called recombination hotspots, are more likely to be involved in crossovers than others. This means that a genetic map distance does not necessarily indicate the physical distance between two markers.

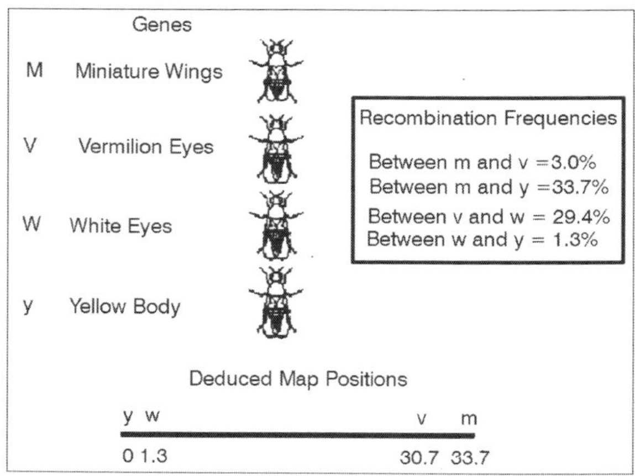

Fig. Working out a Genetic map from Recombination Frequencies.

Also, we now realise that a single chromatid can participate in more than one crossover at the same time, but that there are limitations on how close together these crossovers can be, leading to more inaccuracies in the mapping procedure. Despite these qualifications, linkage analysis usually makes correct deductions about gene order, and distance estimates are sufficiently accurate to generate genetic maps that are of value as frameworks for genome sequencing projects.

LINKAGE ANALYSIS IN DIFFERENT ORGANISM

To see how linkage analysis is actually carried out, we need to consider three quite different situations:

- Linkage analysis with species such as fruit flies and mice, with which we can carry out planned breeding experiments;
- Linkage analysis with humans, with whom we cannot carry out planned experiments but instead make use of family pedigrees;
- Linkage analysis with bacteria, which do not undergo meiosis.

Linkage Analysis in Plant Breeding Experiments

The first type of linkage analysis is the modern counterpart of the method developed by Morgan and his colleagues. The method is based on analysis of the progeny of experimental crosses set up between parents of known genotypes and is, at least in theory, applicable to all eukaryotes. Ethical considerations preclude this approach in humans, and practical problems such as the length of the gestation period and the time taken for the newborn to reach maturity (and hence to participate in subsequent crosses) limit the effectiveness of the method with some animals and plants.

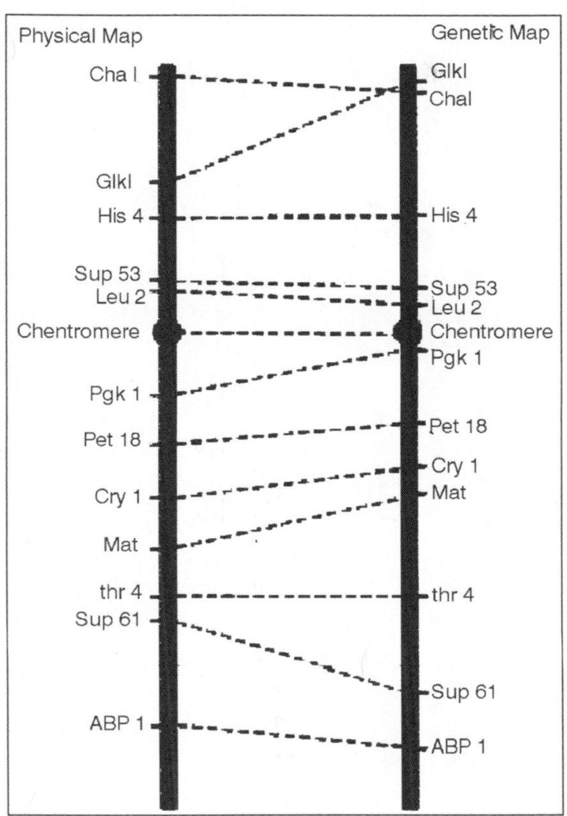

Fig. Comparison Between the Genetic and Physical maps of Saccharomyces Cerevisiae Chromosome III.

If we return to *Figure* we see that the key to gene mapping is being able to determine the genotypes of the gametes resulting from meiosis. In a few situations this is possible by directly examining the gametes. For example, the gametes produced by some microbial eukaryotes, including the yeast *Saccharomyces cerevisiae*, can be grown into colonies of haploid cells, whose genotypes can be determined by biochemical tests. Direct genotyping of gametes is also possible with higher eukaryotes if DNA markers are used, as PCR can be carried out with the DNA from individual spermatozoa, enabling RFLPs, SSLPs and SNPs to be typed. Unfortunately, sperm typing is laborious. Routine linkage analysis with higher eukaryotes is therefore carried out not by examining the gametes directly but by determining the genotypes of the diploid progeny that result from fusion of two gametes, one from each of a pair of parents. In other words, a genetic cross is performed.

The complication with a genetic cross is that the resulting diploid progeny are the product not of one meiosis but of two (one in each parent), and in most organisms crossover events are equally likely to occur during production of the male and female gametes. Somehow we have to be able to disentangle from the genotypes of the diploid progeny the crossover events that occurred in each of these two meioses. This means that the cross has to be set up with care. The standard procedure is to use a test cross. This is illustrated in *Figure,* Scenario 1, where we have set up a test cross to map the two genes we met earlier: gene A (alleles *A* and *a*) and gene B (alleles *B* and *b*), both on chromosome 2 of the fruit fly.

The critical feature of a test cross is the genotypes of the two parents:

- One parent is a double heterozygote. This means that all four alleles are present in this parent: its genotype is *AB*/*ab*. This notation indicates that one pair of the homologous chromosomes has alleles *A* and *B*, and the other has *a* and *b*. Double heterozygotes can be obtained by crossing two pure-breeding strains, for example *AB*/*AB* × ab/*ab*.
- The second parent is a pure-breeding double homozygote. In this parent both homologous copies of chromosome 2 are the same: in the example shown in Scenario 1 both have alleles *a* and *b* and the genotype of the parent is *ab*/*ab*.

The double heterozygote has the same genotype as the cell whose meiosis we followed in *Figure*. Our objective is therefore to infer the genotypes of the gametes produced by this parent and to calculate the fraction that are recombinants. This means that, as shown in Scenario 1 in *Figure*, the genotypes of the diploid progeny can be unambiguously converted into the genotypes of the gametes from the double heterozygous parent.

The test cross therefore enables us to make a direct examination of a single meiosis and hence to calculate a recombination frequency and map distance for the two genes being studied. Just one additional point needs to be considered. If, as in Scenario 1 in Figure, gene markers displaying dominance and recessiveness are used, then the double homozygous parent must have alleles for the two recessive phenotypes; however, if codominant DNA markers are used, then the double homozygous parent can have any combination of homozygous alleles (i.e. AB/AB, Ab/Ab, aB/aB and ab/ab). Scenario 2 in Figure shows the reason for this.

CHROMOSOME NUMBER IN SOMATIC HYBRIDS

The chromosome number in the somatic hybrids is generally more than the total number of both of the parental protoplasts. If the chromosome number in the hybrid is the sum of the chromosomes of the two parental protoplasts, the hybrid is said to be symmetric hybrid. Asymmetric hybrids have abnormal or wide variations in the chromosome number than the exact total of two species.

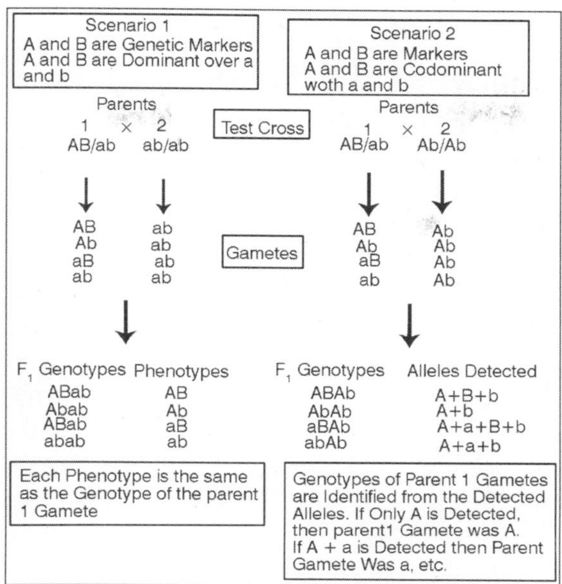

Fig. Two Examples of the test Cross. In Scenario 1, A and B are Genetic Markers with Alleles *A, a, B* and *b*.

In 1972, Carlson and his associates produced the first inter-specific somatic hybrid between Nicotiana glauca and N. langsdorffii. In 1978, Melchers and his co-workers developed the first inter-genetic somatic hybrids between Solanum tuberosum (potato) and Lycopersicon esculentum (tomato). The hybrids are known as 'Pomatoes or Topatoes'.

LIMITATIONS OF SOMATIC HYBRIDISATION

- Somatic hybridisation does not always produce plants that give fertile and visible seeds.
- There is genetic instability associated with protoplast culture.
- There are limitations in the selection methods of hybrids, as many of them are not efficient.
- Somatic hybridisation between two diploids results in the formation of an amphidiploid which is not favourable therefore haploid protoplasts are recommended in somatic hybridisation.
- It is not certain that a specific character will get expressed in somatic hybridisation.
- Regenerated plants obtained from somatic hybridisation are often variable due to somaclonal variations, chromosomal elimination, organelle segregation etc.
- Protoplast fusion between different species/genus is easy, but the production of viable somatic hybrids is not always possible.

Cybrids

The cytoplasmic hybrids where the nucleus is derived from only one parent and the cytoplasm is derived from both the parents are referred to as cybrids. The process of formation of cybrids is called cybridisation. During the process of cybridisation and heterokaryon formation, the nuclei are stimulated to segregate so that one protoplast contributes to the cytoplasm while the other contributes nucleus alone. The irradiation with gamma rays and X-rays and use of metabolic inhibitors makes the protoplasts inactive and non-dividing. Some of the genetic traits in certain plants are cytoplasmically controlled. This includes certain types of male sterility, resistance to certain antibiotics and herbicides. Therefore cybrids are important for the transfer of cytoplasmic male sterility (CMS), antibiotic and herbicide resistance in agriculturally useful plants. Cybrids of Brassica raphanus that contain nucleus of B. napus, chloroplasts of atrazinc resistant B. capestris and male sterility from Raphanus sativas have been developed.

In Vitro Plant Germplasm Conservation

Germplasm refers to the sum total of all the genes present in a crop and its related species. The conservation of germplasm involves the preservation of the genetic diversity of a particular plant or genetic stock for it's use at any time in future. It is important to conserve the endangered plants or else some of the valuable genetic traits present in the existing and primitive plants will be lost. A global organisation- International Board of Plant Genetic Resources (IBPGR) has been established for germplasm conservation and provides necessary support for collection, conservation and utilisation of plant geneic resources through out the world.

The germplasm is preserved by the following two ways:

- *In-situ conservation*: The germplasm is conserved in natural environment by establishing biosphere reserves such as national parks, sanctuaries. This is used in the preservation of land plants in a near natural habitat along with several wild types.
- *Ex-situ conservation*: This method is used for the preservation of germplasm obtained from cultivated and wild plant materials. The genetic material in the form of seeds or in vitro cultures are preserved and stored as gene banks for long term use.

In vivo gene banks have been made to preserve the genetic resources by conventional methods *e.g.* seeds, vegetative propagules, etc. *In vitro* gene banks have been made to preserve the genetic resources by non - conventional methods such as cell and tissue culture methods. This will ensure the availability of valuable germplasm to breeder to develop new and improved varieties.

The methods involved in the in vitro conservation of germplasm are:

- *Cryopreservation*: In cryopreservation (Greek-krayos-frost), the cells are preserved in the frozen state. The germplasm is stored at a very low temperature using solid carbon dioxide (at -79°C), using low temperature deep freezers (at -80°C), using vapour nitrogen (at - 150°C) and liquid nitrogen (at-196°C). The cells stay in completely inactive state and thus can be conserved for long periods. Any tissue from a plant can be used for cryopreservation *e.g.* meristems, embryos, endosperms, ovules, seeds, cultured plant cells, protoplasts, calluses. Certain compounds like- DMSO (dimethyl sulfoxide), glycerol, ethylene, propylene, sucrose, mannose, glucose, praline, acetamide etc are added during the cryopreservation. These are called cryoprotectants and prevent the damage caused to cells (by freezing or thawing) by reducing the freezing point and super cooling point of water.

- *Cold Storage*: Cold storage is a slow growth germplasm conservation method and conserves the germplasm at a low and non-freezing temperature (1-9°C). The growth of the plant material is slowed down in cold storage in contrast to complete stoppage in cryopreservation and thus prevents cryogenic injuries. Long term cold storage is simple, cost effective and yields germplasm with good survival rate. Virus free strawberry plants could be preserved at 10°C for about 6 years. Several grape plants have been stored for over 15 years by using a cold storage at temperature around 9°C and transferring them in the fresh medium every year.
- *Low pressure and low oxygen storage*: In low- pressure storage, the atmospheric pressure surrounding the plant material is reduced and in the low oxygen storage, the oxygen concentration is reduced. The lowered partial pressure reduces the in vitro growth of plants. In the low-oxygen storage, the oxygen concentration is reduced and the partial pressure of oxygen below 50 mmHg reduces plant tissue growth. Due to the reduced availability of O_2, and reduced production of CO_2, the photosynthetic activity is reduced which inhibits the plant tissue growth and dimension. This method has also helped in increasing the shelf life of many fruits, vegetables and flowers.

The germplasm conservation through the conventional methods has several limitations such as short-lived seeds, seed dormancy, seed-borne diseases, and high inputs of cost and labour. The techniques of cryo-preservation (freezing cells and tissues at –196°C) and using cold storages help us to overcome these problems.

X-CHROMOSOME AS TYPE

The chromosome combinations present in the fertilized egg, and this has brought to light some of the methods by which the chromosomes act on development and characteristics. We now Examine other effects of altering chromosomes. We shall have to examine the effects of altering all the different classes of chromosomesýÿthe X-chromosome, the Y-chromosome, and the different autosomes. Each of these three groups (X, Y, and autosomes) gives rise, as we shall see, to a different type of inheritance. The simplest relations, from the experimental point of view, are presented by the X-chromosomes. We therefore deal first with these.

EFFECTS ON ALTERING THE X-CHROMOSOMES

The X-chromosomes take such a course in passing from generation to generation that it is possible to follow the descendants of a particular Xchromosome (that present in the original male parent, for example), knowing in which individuals they are present, in which they are absent. In certain individuals (males, in organisms of Group I) there is but a single X-chromosome instead of a pair. These relations make it a relatively simple matter to discover the distinctive effects of a particular X-chromosome. The effects of X-chromosomes will therefore be dealt with somewhat fully as a type of chromosomal action. Later the action of the other chromosomes will be taken up.

ABNORMAL DISTRIBUTION OF X-CHROMOSOMES

What will happen if the X-chromosomes by accident become irregularly distributed? We know that sometimes the X-chromosomes are indeed irregularly distributed. Will the sex-linked characteristics continue to follow them; will the defective characteristics show the same irregular distribution as do the X's? Such cases in great number have been fully observed. It is found that the sex-linked characteristics do indeed follow the X-chromosomes wherever they go. This proves conclusively (if there were any possible doubt in view of the extraordinary course normally followed by such characters) that it is indeed the X-chromosomes on which the characters depend. The matter is one of importance and interest, so that it will be worth while to examine carefully certain typical cases of the result of irregular distribution of X-chromosomes.

Normal Distributions

Sometimes the body of Drosophila is yellow instead of the normal grey; this is a recessive character due to a defect in the X-chromosomes. Suppose that we mate together a female that has a yellow body and a male that has the normal grey programmes. The female has two modified X-chromosomes, which we may represent by XX. The male has its X unmodified; we may represent its cells as XY. Normally in forming the germ cells, the two X's of the female separate into different germ cells, and all the ova receive one of the modified X's. The normal male produces germ cells of two classes, X and Y.

When the germ cell X from the female unites with Y from the male, sons are produced, with the constitution XY; while X from the female with X from the father gives daughters XX. Since the yellow programmes dependent on X is recessive, the daughters XX have the normal grey programmes, like the father. But the sons XY have only one X, and as this is the defective one, from the mother, the sons are yellow, like the mother. The normal result of such a mating therefore is that all the sons are recessive like the mother, all the daughters dominant, normal, like the father. This is what occurs when the X-chromosomes are distributed to the germ cells in the usual way.

ABNORMAL DISTRIBUTION

In forming germ cells both X's of the female go together to a single ovum, while other ova receive no X. In a race discovered by L. V. Morgan the two X-chromosomes in the female were partly united, so that they almost invariably thus go together to one ovum, leaving other ova without an X. In this race the X-chromosomes were so modified as to produce the recessive yellow programmes described in the preceding paragraph. We therefore have an opportunity to determine the course of inheritance of the yellow and the grey programmes when the X-chromosomes are thus abnormally distributed. Represent as before the X-chromosomes that produce the yellow programmes by X. Thus one set of ova receive XX, while another set receive no X.

When the germ cells XX united with the germ cells Y from the normal father, there were produced daughters XXY, in which both X's came from the mother, and these daughters were yellow like the mother (since no normal X was present). Sons were produced by the union of ova that contained no X with sperms that carried X; this gave sons XO, with their single X from the normal father, instead of from the mother, as is usually the case. And such sons had the normal grey body programmes of the father, instead of the yellow body programmes of the mother, as happens normally.

Thus with a change in the distribution of the X's there is a corresponding change in the method of inheritance. If the sons receive their X from the recessive mother, as in the normal cases, they are recessive like the mother. But when, through non-disjunction, the sons receive a dominant X from the father, they are dominant like the father. Similarly, in the normal case the daughters receive a dominant X from the father and are therefore dominant like the father. But in cases of non-disjunction the daughters receive both their JPs from the recessive mother, and are therefore recessive like the mother.

In sum, when the X's are normally distributed, such matings give 'criss-cross inheritance'; sons like the mother, daughters like the father. But when the X's are not so distributed, there is no criss-cross inheritance; sons are like the father, daughters like the mother. The dominant and recessive characters follow the respective X-chromosomes, whether these are distributed normally or abnormally. Such experiments have been repeated many times, and with other sex-linked characters. Always the characteristics follow the distribution of the X-chromosomes, however these are distributed. It may be concluded with certainty that it is the presence of X-chromosomes of a modified type that causes the appearance of the particular sex-linked characters that are manifested. The rules of distribution of sex-linked characters are the rules of distribution of the X-chromosomes.

DOMINANCE AND RECESSIVENESS IN SEX-LINKED CHARACTERS

As we have seen, and shall see further, many defective conditions are due to defects in certain X-chromosomes. These follow from generation to generation the distribution of these X-chromosomes. Most of these bodily defects are manifested only in individuals in which the defective X-chromosomes are the only kind present; that is, these defects are recessive. But it is important to observe that in such cases the normal condition of the organism likewise follows the distribution of certain X-chromosomes.

Haemophilia follows certain X- chromosomes. But in the same matings in which this occurs, the healthy condition of the blood follows certain other X-chromosomes. If a defective mother is mated with a normal father, the sons are defective because they receive only the mother's defective X-chromosomes. But in the same way the daughters are normal because they receive the father's normal X-chromosomes. In the normal individuals the X-chromosome plays a part, just as it does in the defective individuals. In the normal individuals it is healthy and supplies what is required for normal development, while in the defective individuals it fails to supply what is required. In the case of most defects this normal condition is dominant over the recessive condition; that is, when both are present, the normal chromosome is the one that prevails in its effect on the individual. There are defective characters, however, in which the defective condition is dominant or partly dominant.

In these cases, when a defective X and a normal one are present together, the defective one produces its effect, as in the case of bar-eye, described earlier. In many such cases the defect is less marked when a normal chromosome is present as well as a defective one; then the defective condition is said to be partly dominant. Thus whenever an individual bearing defective X-chromosomes is mated with a normal individual, we have both dominant and recessive characters, the course of which may be followed in the later generations the dominant character being commonly the normal or usual condition.

TESTS FOR SEX-LINKED INHERITANCE

By observing the distribution of the dominant and recessive characters among the offspring of certain matings, a test for determining whether given characters depend on the X-chromosomes is supplied. The test consists in making what are called reciprocal crosses.

- On the one hand mate a dominant female with a recessive male;
- Also mate a recessive female with a dominant male.

Represent a dominant X-chromosome by a capital X, a recessive X-chromosome by a lower-case letter x. In the group of organisms with which we have been dealing, the female has two X-chromosomes, the male but one. The daughters receive an X from each parent, the sons an X from the mother only. A dominant female will be represented by XX, a recessive male by xo; similarly a recessive female is xx, a dominant male is XO. The two ma tings and their results will then be represented as follows:

- XX by xo gives X + XO Dom. Mother Rec. Father Dom. Daughters Dom. Sons
- xx by XO gives X + xo Rec. Mother Dom. Father Dom. Daughters Rec. Sons

Thus when the mother is dominant, the father recessive, all the children are dominant, like the mother. When the father is dominant, the mother recessive, the daughters are dominant like the father, the sons recessive like the mother; this is called 'criss-cross inheritance'.

Whenever reciprocal crosses give these results, it is certain that the two diverse characteristics (dominant and recessive) result from differences in the X-chromosomes of the two parents. 'Criss-cross inheritance' is particularly useful in showing at once that we are dealing with sex-linked inheritance. In every case where such results are produced, further tests show that the two characteristics follow in later generations the two different kinds of X-chromosomes wherever they go. It is mainly by the use of these tests that the many different characteristics dependent on diversities in X-chromosomes have been discovered.

SEX-LINKED INHERITANCE IN GROUP II

Sex-linked inheritance as we have just described it was originally discovered in animals belonging to Group I, in which the females have two X-chromosomes, the males but one (with or without a Y- chromosome). But when the tests we have just described were applied to birds and to certain other animals, it was discovered that there is a second group (Group II) in which it is the male that has two X-chromosomes, the female but one. In the common fowl, reciprocal crosses were made between 'barred' fowls (Plymouth Rock) and black fowls (Langshan). Here the 'barred condition was found to be dominant.

The results were:

- Barred father by black mother=Barred sons + barred daughters.
- Black father by barred mother = Barred sons + black daughters.

Here the second mating gives 'criss-cross inheritance', showing that we are dealing with sex-linked characters dependent on differences between the X-chromosomes of the parents. But, in the first mating above, all the offspring are like the father instead of like the mother, which latter was the case in the matings. The father in this case has the dominant characteristic (barred), and all the offspring are dominant like the father instead of like the mother.

When the mother is dominant ('barred') the result in the fowls is criss-cross inheritance; while, in the cases before described, criss-cross inheritance occurs when it is the father that is dominant. In fact, in the birds, the males and females simply exchange roles, as compared with their roles in the organisms of Group I.

Since the roles of males and females are interchanged in the two groups, their chromosomal conditions must also be interchanged; that is, since in Group I the female has two X's, and in Group II the male plays the same role as the female of Group I, the male of Group II must have two X's, the female but one. The chromosomal conditions in the two groups are shown in the above tabulation, X signifying dominant, while x signifies recessive, and o signifies the lack of X (whether Y is present or not).

Careful examination shows that this is the only way in which the results in Group II can be produced. Birds and certain moths show inheritance of the kind typical for Group II. In both these it has since been found under the microscope that the males have indeed one more chromosome than the females.

ROLE OF DEVELOPMENT IN X-CHROMOSOMES

A great many characteristics, of various kinds, have been found to be the result of modifications of particular X-chromosomes, and thus to follow in later generations the distribution of the descendants of those chromosomes. Examination of a number of these is desirable, both for their own importance and for the light they throw on the functions of the X-chromosomes.

In man, the following characteristics, among others, dependent on diverse types of X-chromosomes, are known from the fact that they show typical sex-linked inheritance.

Haemophilia: Lack of coagulability in the blood. This results from a serious defect in certain X-chromosomes. The existence of defective X-chromosomes having this result shows that the normal X-chromosomes play a part in supplying something necessary for producing normal blood that coagulates properly.

Programmes Blindness: The fact that defectiveness in X-chromosomes causes programmes blindness shows that the normal X's play a role in producing the normally functioning eyes.

NIGHT BLINDNESS

- Inability to see in a poor light.
- Near-Sightedness of certain types.

Progressive atrophy of the muscles. The fact that defects in X-chromosomes have this effect shows that the normal X's play a role in the normal functioning of the nerves and muscles. A considerable number of other sex-linked characters are known in man. But man is a very unfavourable organism for the study of inheritance. Yet even the little that is known of sex-linked inheritance in man shows that the X-chromosome plays a role in many diverse bodily functions.

For a more complete idea of its role, some organism that can be bred experimentally must be examined. For this purpose the fruit-fly, Drosophila melanogaster, is the best organism to select, since it has been studied more extensively than any other. In the fruit-fly we find peculiarities of the following types that are dependent on alterations in particular X-chromosomes:

Many different eye colours: The normal eye programmes in this organism is a certain shade of red. Defects in different X-chromosomes result in producing, in the individuals that bear these defective chromosomes, many different shades of red, varying from deep red to a very light red, and thence to 'buff', 'ivory, and 'white'. More than a dozen different types of eye programmes are known to result from modifications of the X-chromosome. It is clear that the normal X-chromosome plays an important role in producing the normal eye programmes.

Structural peculiarities of the eye, such as 'bar-eye', 'facet eye', 'furrowed eye', and the like, are known to result from modifications of the normal X-chromosome.

- *Wing modifications*: Many different conditions of the wings are known that depend upon diversities among X- chromosomes borne by different individuals. These affect all sorts of features of the wings: size, form, venation, function. The normal X-chromosome obviously plays an important role in the full and normal development of the wings.
- *Body colours and markings*: Modifications of the X- chromosomes produce the body colours yellow, sable, tan, chrome, lemon, green, and the like, in place of the normal grey. Other changes in X alter the distribution of pigment on the body, giving the characters dot.
- *Body structure*: A defect in X produces irregularities in the abdomen, known as 'abnormal abdomen'.
- *Legs*: A defect in X-chromosomes causes abnormal development of the legs; some of them are wholly or partly reduplicated, so that the total number of legs is increased.
- *Bristles*: Modifications in X-chromosomes result in various different changes in the bristles that are scattered over the body.

Many physiological conditions and functions likewise depend upon the X-chromosomes, since they are changed when these chromosomes are modified.

Among these are the following:

- *Positive reaction to light*: Drosophila with normal X-chromosomes fly towards a source of light. Those having defective X-chromosomes of the kind that produce a tan-coloured bodydo not fly towards a source of light.
- *Weakness and short life*: Most of the, defective X-chromosomes that cause structural or other changes in the body (different eye colours, wing forms, and the like) produce likewise weakness and short life. The individuals bearing them are less resistant to bad conditions, and live for a shorter time than the individuals that bear the normal X-chromosomes.
- *Life and death*: Some X-chromosomes have defects that are so severe that the individuals bearing them will not live and develop, unless there is present also a normal X-chromosome. Such defects are known as lethals. The presence of such lethal defects in the X-chromosome has the result that males bearing them die, since they have but one X; while females usually live, since they have an additional X that is often normal.

If a mother that has one lethal X and one normal one is mated to a father that has a normal X, the result is that half of the sons which receive the mother's defective X fail to develop, while the other half of the sons,receiving the mother's normal X, live and develop. All the daughters live, since they all receive a normal X-chromosome from the father. The consequence is that in such families there are twice as many daughters as there are sons. In families of 100 to 200, as occur in Drosophila, this is very striking. X-chromosomes are known to play in other organisms roles similar to those mentioned above for Drosophila, although in no other organism has the matter been so fully studied. *We may summarize what has been brought out above in the following statements*:

- Many defects that appear in individuals are the result of defects in the X-chromosomes that they bear.
- Some of these defects are dominant; they are manifested in all individuals in which the defective X is present.
- But most such defects are recessive; they are manifested only in individuals in which the defective X is the only kind present.
- When both a normal X and a defective X are present, in most cases the normal X performs the required functions, so that the individual is not defective.
- Since the female has two X-chromosomes, while the male has but one, such defects are more frequently manifested in the males. Usually one of the X's present in the female is without the defect, so that she is not defective. But when the male carries a defective X-chromosome, the defect is manifested.

- It is thus advantageous, so far as the occurrence of defects is concerned, to have two X-chromosomes rather than one.
- This gives the female a considerable advantage over the male in these respects. There are many different types of defects due to defective X-chromosomes; almost all of them are more common in males than in females. Some of these defects seriously injure the health, or even result in death, if the defective X is the only kind that is present. It is well known that, in general, the death rate is higher in males than in females; this is probably due to the fact that males have but one X-chromosome.
- Among the different individuals of a species are scattered a great number of different types of X-chromosomes having different effects on development. Many of the diverse X-chromosomes are distinctly defective, causing personal defects in the individuals that bear them. Others are not defective.
- Defects or modifications in X-chromosomes affect in different cases all parts and functions of the organism

What do the facts brought out in the preceding sections show as to the physiological functions of the X-chromosomes; as to their role in development? For every kind of effect resulting from defects or modificacations in the X-chromosomes, there is a corresponding (opposed or diverse) action of normal or unmodified X-chromosomes. Since defective X-chromosomes of a certain type produce in Drosophila short, ill-developed wings, it follows that the normal X-chromosomes act on the development in such a way as to give long, well-developed wings.For if we substitute the normal X for the defective one, this causes the normal well-developed wings to be produced.

Similar reasoning applies to all the defective conditions that result from defective X's; the normal X's act in such a way as to produce the corresponding normal conditions. It follows therefore that the normal X-chromosomes play a role in the production of eye programmes and structure, body programmes and structure, in the development of the legs, the wings, the bristles, in producing normal health, resistance and vigour, and in various physiological processes. We know, further, that they play a most important part in determining sex, with all that this includes of structural and physiological influence. Clearly therefore the X-chromosomes play a most important part in development. Their action is not limited to any single part of the body, nor to any single class of functions. They enter into the processes of development in such a way as to influence all parts of the body, all functions of the body. They begin to influence development very early, as was seen in considering the development of sex. And they play important roles in such relatively late processes as the production of eye colours. It can hardly be doubted that they enter into the developmental activities from practically the beginning, influencing all the later processes that occur.

AUTOSOMES AND T-CHROMOSOMES

The X-chromosomes have been dealt with because of the great advantages they offer for experimental study. It was seen that they affect development in many ways, and that in a given species there are many different types of X-chromosomes in the different individuals, causing them to show different characteristics, structural and physiological.

Is the X-chromosome typical in these relations? Shall we find similar effects in the other chromosomes? In addition to the X-chromosomes, there are the autosomes, which are present as a rule as several or many pairs. In man there are 23 pairs of autosomes: in Drosophila there are three pairs. There is, further, in many species the single Y-chromosome. The autosomes and the Y-chromosome are distributed from parent to offspring in characteristic ways, differing from the distribution of X-chromosomes. If they affect characteristics, these differences should appear in a different system of inheritance. We shall examine first the autosomes.

THE AUTOSOMES: TYPICAL MENDELIAN INHERITANCE

We have already seen that the autosomes play a role in development, in the fact that they help to determine the sex of the developing individual. In Drosophila, if the number of sets of autosomes, in relation to the number of X's, is changed, this changes the sex.Do autosomes also affect other characteristics? Do the different pairs of autosomes have different functions? Are there diverse types of autosomes, with different effects, in different individuals of a species, as is the case in relation to the X-chromosomes? Are there dominant and recessive effects of autosomes, as there are of X?

If dominant and recessive characteristics depend on autosomes, then these characteristics must follow the distribution of the autosomes from parents to offspring, as the characteristics dependent on X follow its distribution. We must then examine the method of distribution of the autosomes whether there are characteristics that show this method.

11

Forest Trees and Plantations

TREES OUTSIDE THE FOREST

The significance of trees outside the forest (TOF) can be observed in several contexts. In countries with low forest cover, TOF resources constitute the main source of wood and non-wood "forest" products, even though trees may be so scattered that the maps produced by FRA 2000 indicate that no forests exist. Trees are found on agricultural lands, in densely populated areas, in fruit-tree plantations and in home gardens, which often cover a large proportion of the land. In urban areas trees provide important aesthetic and environmental services in addition to providing shade and greatly increasing the livability of cities. Communities, farmers and herders who do not have access to forests diversify their production and protect their land by maintaining various tree systems on their farms.

Deforestation has been mapped and quantified, but very little is known about the fate of land formerly under forest; forest clearing is often followed by the establishment of production systems of which trees are an integral part. Not much is known about the dynamics of trees on farmlands and their corresponding contribution to the production of wood and other products and services. Similarly, little is known about changes in tree cover in fields and urban systems. Knowledge of trees outside the forest comes mostly from local studies on agroforestry, sylvipastoralism and urban, social, community or rural forestry. This widespread and multipurpose resource, familiar to farmers but poorly defined by managers and mostly absent from official statistics and development policies, needs to be better assessed and known.

Growing populations, shrinking forests and degraded ecosystems all suggest that trees outside the forest are destined to play a larger local and global role in meeting the challenges of resource sustainability, poverty reduction and food security. Trees outside the forest relieve the pressure on forest resources, conserve farmland, boost agricultural productivity, blunt the harmful impact of urban growth on the environment, increase food supplies, provide income and in general make valuable contributions to food security.

FRA 2000 did not undertake a global assessment of trees outside the forest, mainly because of resource limitations; nor has there ever been a comprehensive global assessment of trees outside the forest and their products. However, a number of studies have been carried out for specific sectors or geographic areas, often with an emphasis on their economic contributions. This chapter provides a summary of selected studies and discusses the practical and conceptual difficulties related to a comprehensive global assessment. The chapter responds to the concern expressed by the Expert Consultation on Forest Resources Assessment 2000 regarding the lack of information on TOF.

METHODS AND TOOLS FOR FUTURE ASSESSMENTS

A priority challenge of future assessments is to know the state and dynamics of all tree resources both in and outside the forest. A country embarking on a planning exercise cannot confine itself solely to the trees within its forests, especially when its wood resources appear to be insufficient. The choice of tools and methods used to describe or assess trees outside the forest depends on the scale of analysis, kind of data and degree of exactitude desired. The tools used are not generally specific or new; rather, they are combined and implemented in original ways.

The inventory in Bangladesh described above is one of the numerous examples of methods developed for gathering data on TOF. The Bangladesh study gives evidence of the adaptations needed to bring conventional forest inventory procedures in line with the specific nature of this resource. In some aspects - e.g. structure, spatial distribution and extent of area cover - trees outside the forest are more difficult to assess than forest formations. The assessment of TOF does not lend itself to the potential cost savings associated with expanded uses of remote sensing technology. Remote sensing by satellite presents more difficulties for assessing TOF resources than for assessing attributes such as forest area. However, satellite data do allow a region to be stratified on the basis of ecological criteria and land cover, providing the basis for a good working document for more specific work in the future. The most commonly used remote sensing technology for TOF resources is aerial photography, which can be used to describe spatial distribution and to distinguish TOF cover classifications, providing the appropriate scale is chosen. However, high costs prohibit widespread use of aerial photography for TOF assessments in most countries. The new 1 m resolution satellite sensors represent a possible future alternative to aerial photography.

Some TOF field inventories are modelled on forest inventory methods and keep to biological and physical criteria; others emphasize social aspects, choosing villages as the sampling units. For measurements on the ground, sampling arrangements designed for forest stands may not be the most effective arrangements for trees. Less traditional sampling plans which would theoretically be better suited to this resource should be tested on various categories of TOF, especially those covering fairly large areas.

Studies of the social and economic benefits or impacts of TOF often rely on household surveys, interviews or standardized appraisals such as rapid or participatory rural appraisal. The integration of the last two approaches - biophysical inventory and socio-economic analysis - is not simple and calls for caution given the great variety of social situations that are only meaningful in the local context. Environmental benefits or impacts of TOF might be indirectly assessed by linking measurable indicators, such as the number and type of trees, with environmental variables such as water quality or erosion. In an urban setting, tree cover might have direct impact on the ambient temperature. Measuring the environmental impact of tree management is an issue for all natural resource planning or management operations.

Assessment of trees outside the forest requires geographical, ecological, biophysical, social and economic data. However, this implies that an important amount of information will have to be carefully processed. The diversity of end-uses for this information, including land use planning and analysis based on inventory, will need to be considered in data assembly and processing and in the presentation of results. It is important to know the status of trees outside the forest at any given moment, but it is even more essential to be able to trace patterns of change over time in the same area. The two most commonly used approaches have been comparison of aerial photos taken at sufficiently long intervals and surveys among villagers/managers combined with field inventories.

Some countries, such as France and the United Kingdom, have undertaken periodic inventories based on the establishment of permanent plots linked to permanent forest inventories. However, the high cost of this type of operation limits the number of countries able to adopt it. India and Bangladesh are now experimenting with options for the future. The current trend towards decentralized authority in land use planning suggests the importance of carrying out assessments at the local level, where the geographical, historical and socio-economic context is relatively harmonious. A minimum number of common rules concerning methods and arrangements is necessary, however, if the data are to be comparable at the country level. Certainly, the technical side of assessing trees outside the forest is complex and more research is needed to better pinpoint the resource.

FOREST PLANTATIONS

Forest plantations covered 187 million hectares in 2000, of which Asia accounted for 62 per cent. The forest plantation area represents a significant increase from the 1995 estimate of 124 million hectares. The reported new annual planting rate is 4.5 million hectares globally, with Asia and South America accounting for 89 per cent. About 3 million hectares are estimated to be successful. Globally, half the forest plantation estate is for industrial end-use, one-quarter for non-industrial end-use and one-quarter not specified. Globally, the main fast-growing, short-rotation species are in the genera *Eucalyptus* and *Acacia*. Pines and other coniferous species are the main medium-rotation utility species, primarily in the temperate and boreal zones.

The potential for forest plantations to partially meet demand from natural forests for wood and fibre for industrial uses is increasing. Although accounting for only 5 per cent of global forest cover, forest plantations were estimated in the year 2000 to supply about 35 per cent of global roundwood. This figure is anticipated to increase to 44 per cent by 2020. In some countries forest plantation production already contributes the majority of industrial wood supply. There is increasing interest in development of forest plantations as carbon sinks; however, failure to resolve international debates on legal instruments, mechanisms and monitoring remains a serious constraint.

In developing countries about one-third of the total plantation estate was primarily grown for woodfuel in 1995 - although it should be noted that planted trees on farmland, in villages and homesteads and along roads and waterways contribute significantly to fuelwood supplies, enabling the demand to be met in most instances. New forest plantation areas were reported as being established globally at the rate of 4.5 million hectares per year, with Asia and South America accounting for more new plantations than the other regions. Of plantations established, about 3 million hectares per year were estimated as being successful. Of the estimated 187 million hectares of plantations worldwide in 2000, Asia had by far the largest area. In terms of composition, *Pinus* spp. (20 per cent) and *Eucalyptus* spp. (10 per cent) remain dominant worldwide, although the overall diversity of species planted was shown to be increasing. Industrial plantations account for 48 per cent, non-industrial plantations for 26 per cent and plantations for unspecified use for 26 per cent of the global forest plantation estate.

The results of the plantation assessment were the first global estimates with a uniform definition of forest plantations and can therefore not be directly compared to previous estimates. FRA 2000 country statistics on plantations may also differ from those reported in prior FAO publications, partly because of changes in definitions. Countries participated directly in the assessment, providing technical documentation and supporting analysis and validating the results generated by FAO. Several experts around the world were enlisted to provide detailed information on various aspects of the plantation situation in the form of special studies.

Between the extremes of afforestation and unaided natural regeneration of natural forests, there is a range of forest conditions in which human interventions occur. European forests have long traditions of human intervention in site preparation, tree establishment, silviculture and protection; yet these are not always defined as forest plantations. The traditional forest plantation concept tends to be applied to single species, uniform planting densities and even age classes. Terms such as "natural forest under management" or "assisted natural regeneration" are applied to stands of indigenous species in more heterogeneous management mechanisms in Europe and other industrialized temperate and boreal countries.

METHODS

The area of existing forest plantations would ideally all have been derived from statistically designed inventories of forest plantations or statistics for planted areas reported by planting agencies or appearing in national reports. However, information also comes from many other sources including nursery production, seedling distribution and estimates derived from the goals of planting programmes.

The vast range of agencies, industries and non-governmental organizations within countries engaged in planting programmes made the comprehensive collection of all relevant source documents a major logistical exercise. For FRA 2000, over 800 source documents were analysed to derive the forest plantation estimates. In most developing countries a national clearinghouse for collecting information on plantations is either lacking or ineffective owing to the enormity of the task and limited resources.

Data Collection

To retrieve the source documents for the plantation study, FAO made formal requests to all developing countries, some of which contributed the necessary materials. Most of the reports were collected directly by FAO staff during FRA 2000 workshops and visits to national ministries. For consistency FRA 2000 prepared guidelines and questionnaires for the collection of forest plantation statistics in which the objectives, scope, definitions, sources of data and templates for specific data collection were supplied to each country. Parameters requested included:

- Total estimated forest plantation area, 2000;
- Annual area of new plantations;
- Species groups: broadleaf, conifer, non-forest like, African oil palm (*Elaeis guineensis*), coconut palm (*Cocos nucifera*) bamboo or unspecified;
- Purpose and end-use objective of forest plantations: industrial (producing wood or fibre for supply to wood processing industry) or non-industrial (fuelwood, soil and water protection);
- Ownership: public, private, other (e.g. traditional, customary) or unspecified.

Other data requested in the guidelines, which proved difficult for countries to provide by species group, included age class distribution; end-use by forest product (industrial plantations); growth and yield (mean annual increment); standing volumes; and rotation lengths. Despite the absence of these data, FRA 2000 is the most comprehensive forest plantation resources assessment that has been carried out.

In previous assessments of forest plantation resources, plantation data were available up to the reference year for most countries, since the reporting followed the reference year. In FRA 2000, the reference year was 2000, so if data were not available to that date, then existing area and annual planting data were used to extrapolate the necessary information. For the few countries that have no data sets since 1990, the rate of planting in preceding years and future planting programmes were considered in projections to the year 2000.

FAO also enlisted the assistance of several experts around the world to make specific technical contributions on the forest plantation situation in the 1990s. These studies constituted an important part of the global results as well and complemented the country information.

Analysis and Interpretation

The quantity and quality of forest plantation data provided is dependent upon the capacity of the national forest inventory systems to collect and analyse data and to adjust the information to conform with global and regional reporting parameters. In many developing countries there is a lack of institutional capacity to carry out periodic national forest inventories, so data can be incomplete, inconsistent, outdated and of variable reliability. Because of this, it was necessary to derive and in some instances to verify forest plantation statistics through desk research using available country reports. All sources of country data were referenced and made available in a transparent manner.

In addition, regional and national focal persons were appointed to assist in the forest plantation data collection, to ensure that the latest data were available and to maintain coordination and communication between FRA 2000, FAO regional offices and each participating country. On completion of the data sets, a formal verification process was undertaken with each participating country.

Purpose and ownership within the global forest plantation estate Purpose and ownership of forest plantations vary markedly among regions. Industrial plantations provide the raw material for wood processing for commercial purposes, including timber for construction, panel products and furniture and pulpwood for paper. In contrast, non-industrial plantations are aimed for example at supplying fuelwood, providing soil and water conservation, wind protection, biological diversity conservation and other non-commercial purposes.

In many countries, particularly in the developing world, the end purpose of the plantations is not clearly defined at the outset. In some of these cases, valuable tree resources are established which coincidentally match future needs. However, in other cases the lack of planning may result in plantations that have little commercial value and a low potential for local use.

SELECTED GLOBAL TRENDS, 1980-2000

Comparisons

FRA 2000 country statistics on plantations may differ from those reported in prior FAO publications partly because of changes in definitions. For example, rubber plantations were not previously considered as forest plantations but are included in FRA 2000 plantation data. Previous assessments also used regional reduction factors to indicate the successful proportion of plantations remaining after establishment. The FRA 2000 assessment applied reduction factors according to the best available data from each country independently. There have also been changes in the information base from which the estimates were derived. The statistics now include data from many industrialized countries, none of which were included in the prior global assessment reports. Despite these differences, comparison of FRA results from each decade allows analysis of some trends including planting rates, genera, areas and purpose.

IMPACTS OF THE FOREST PLANTATION ESTATE

The potential for forest plantations to partially meet demand for wood and fibre for industrial uses is increasing. According to FRA 2000, the global forest plantation area accounts for only 5 per cent of global forest cover and the industrial forest plantation estate for less than 3 per cent. However, as an indication only, forest plantations were estimated in the year 2000 to supply about 35 per cent of global roundwood and an increase to 44 per cent anticipated by 2020. If plantation development is targeted at the most appropriate ecological zones and if sustainable forest management principles are applied, forest plantations can provide a critical substitute for natural forest raw material supply. In several countries industrial wood production from forest plantations has significantly substituted for wood supply from natural forest resources. Forest plantations in New Zealand met 99 per cent of the country's needs for industrial roundwood in 1997; the corresponding figure in Chile was 84 per cent, Brazil 62 per cent and Zambia and Zimbabwe 50 per cent each. This substitution by forest plantations may help reduce logging pressure on natural forests in areas in which unsustainable harvesting of wood is a major cause of forest degradation and where logging roads facilitate access that may lead to deforestation.

Forest plantations also provide additional non-wood forest products, from the trees planted or from other elements of the ecosystem that they help to create. They contribute environmental, social and economic benefits. Forest plantations are used in combating desertification, absorbing carbon to offset carbon emissions, protecting soil and water, rehabilitating lands exhausted from other land uses, providing rural employment and, if planned effectively, diversifying the rural landscape and maintaining biodiversity.

Not all forest plantation development has positive economic, environmental, social or cultural impacts. Without adequate planning and without appropriate management, forest plantations may be grown in the wrong sites, with the wrong species/provenances, by the wrong growers, for the wrong reasons.

Examples exist where natural forests have been cleared to establish forest plantation development or where customary owners of traditional lands may have been alienated from their sources of food, medicine and livelihoods. In some instances poor site/species matching and inadequate silviculture have resulted in poor growth, hygiene, volume yields and economic returns. In other instances, changes in soil and water status have caused problems for local communities. Land use conflicts can occur between forest plantation development and other sectors, particularly the agricultural sector.

The negative impacts of forest plantations can draw the focus away from the fact that forest plantation resources are totally renewable and can be economically, socially, culturally and environmentally sustainable with prudent planning, management, utilization and marketing.

SELECTED FOREST PLANTATION TOPICS

Mean Annual Volume Increment (MAI) of Select Industrial Species

On average *Eucalyptus* and *Pinus* species, which dominate industrial plantations in developing countries, have similar MAIs of 10 to 20m^3 per hectare per year. However, many of the popular species of both genera frequently achieve much faster growth rates. Thus *Eucalyptus grandis,* which is the most widely planted *Eucalyptus* species, can achieve 40 to 50 m^3 per hectare per year and in very exceptional conditions with advanced tree improvement 100m^3 per hectare per year. Other widely planted tropical hardwoods including *Casuarina equisetifolia, Casuarina junghuhniana*, *Tectona grandis* and *Dalbergia sissoo* have MAIs of less than 15m^3 per hectare per year and frequently under 10 m^3 per hectare per year.

Climate and site have a very large impact on growth rates. The humid tropics and more fertile sites are more conducive to higher growth rates than locations with long dry seasons or infertile or degraded soil. Teak on many sites in India, for example, frequently has an MAI of 4 to 8 m^3 per hectare per year, partly because of drought combined with poor soils. Some species such as *Gmelina arborea* and some of the *Eucalyptus* species are very site sensitive. *Pinus* spp., in contrast, generally tolerate adverse conditions better and are more flexible with respect to site. Both tree breeding and silviculture have improved growth rates. Good examples are *Eucalyptus grandis* and *E. urophylla* in Brazil and *Pinus radiata* in some countries of the Southern Hemisphere.

Advanced silviculture typically includes improved nursery and establishment techniques such as good site preparation, weed control and judicious use of fertilizer. It has been suggested that growth of teak, for example, could be doubled in Kerala, India and Bangladesh and increased sixfold in Indonesia by adopting these practices.

With coppice species productivity varies with rotation, the first and second coppice rotations usually being more productive than the seedling one. The growth patterns vary among species. For example, very fast growing species such as *Gmelina arborea* can reach a peak MAI in less than 10 years, while *Pinus caribaea* var. *hondurensis* grown in Trinidad reaches maximum MAI at about 25 years and *P. radiata* at over 40 years. With *Cupressus lusitanica* in Costa Rica, the MAI maximum is reached at about 30 years.

Rotation lengths can reflect both end-use and economics. Many fast-growing *Eucalyptus, Acacia* and *Casuarina* species and *Gmelina arborea* are grown on short rotations of under 15 years as they are used primarily for pulp or woodfuel. Usual rotations in Kenya for *E. grandis* are 6 years for domestic woodfuel, 7 to 8 years for telephone poles and 10 to 12 years for industrial woodfuel. In Brazil this species is largely grown for pulp or charcoal on 5 to 10 year rotations. Species being grown for high-value sawlogs usually have longer rotations; teak (*Tectona grandis*) is grown on 50 to 70 year rotations and high-value conifers such as *Araucaria angustifolia* on 40 year rotations. Generally pines are grown on medium-length rotations of 20 to 30 years, unless grown solely for pulpwood, when shorter rotations may be adopted.

Modelling of growth, rotations, harvest yields and product mix by species is important for decision-making in forest management. One of the major obstacles to model development for planners and managers is a lack of suitable data. Data can come from a range of sources, including temporary and permanent sample plots and experiments. Experiments and protocols for obtaining data need to be carefully designed so that reliable information is obtained over the complete range of conditions to which the model is to apply. Tree Growth and Permanent Plot Information System (TROPIS), sponsored by the Centre for International Forestry Research (CIFOR), seeks to coordinate and improve access to tree growth information.

A growth and yield model developed for *Pinus elliottii* plantations in coastal Zululand, South Africa, predicts height, basal area, total stem and merchantable volume and stocking, with age on a stand level for harvest planning. In New Zealand several simulation models have been developed for *P. radiata* which predict similar variables but also include wood quality, harvesting and marketing aspects, which make it possible to link the silvicultural options to industrial use.

Sustaining Productivity

It is possible not only to sustain but also to increase productivity in successive rotations. This requires clear definition of the end-use objective for forest plantation development and a holistic view in their management. There is a need to integrate strategies for genetic improvement programmes, nursery practices, site and species/provenance matching, appropriate silviculture (site preparation, establishment, weeding, fertilizing, pruning, thinning), forest protection and harvesting practices with prudent management. New Zealand and the southern United States have shown that substantial gains can be made by adopting this holistic approach.

In developing countries where resources may be constrained, highly technical solutions may not be essential but it is critical to get the fundamentals correct: careful species and provenance choice, good nursery stock, site preparation, planting techniques, weed control and, less frequently, fertilizer inputs. Once fast-growing, uniform plantations have been established, later silvicultural tending may become increasingly important, depending on the end-use objective.

Current evidence suggests that plantation production can be sustainable if foresters implement prudent genetic and silvicultural tree improvement programmes and sound management practices. There has been, however, limited long-term research on the subject; there are few definitive studies, limited to few species.

In one of the most promising studies, with *Pinus patula* in Swaziland grown intensively on about 15 year rotations, site productivity was maintained or increased over three rotations. The question of declining growth in teak (*Tectona grandis*) plantations in Indonesia and India remains unclear.

How forest plantations are managed affects the chemical and physical properties of the soils and site. However, only recently have long-term studies been undertaken to evaluate these critical factors or processes.

The methods adopted for site preparation (ripping, ploughing, scarifying, bedding, windrowing, controlled burning), establishment (manual, mechanical), weeding (manual, chemical, mechanical), fertilizer application, pruning and thinning (manual and mechanical, for commercial to waste), forest protection and harvesting (manual, mechanical, clear-fell or selection) all affect the pool of nutrients in the ecosystem. Interference with the drainage, litter and recycling of organic matter and change in the physical conditions of soils during these operations are critical to long-term sustainability. Because of litter recycling and the rapid development of tree roots, plantations are used for rehabilitation of fragile and degraded lands prone to soil erosion and excessive water runoff.

Tree plantations often have higher evapotranspiration rates than grassland or agricultural crops and thus change the hydrology of the site. This can be either beneficial (for example by reducing salinity problems in some dryland conditions) or detrimental (if it reduces water required for other uses). The rare studies of changes of productivity between rotations have concluded that negative changes have primarily been due to inappropriate or inadequate management practices or weed invasions rather than a result of the plantations themselves.

Burning and excessive cultivation in site preparation, soil compaction from mechanical operations, inappropriate harvesting techniques and poor forest protection can contribute to loss of nutrients and soil erosion, with a resultant loss in productivity of forest plantation sites. This cannot be addressed solely by addition of fertilizer, but by the adoption of the whole range of tree improvement, silviculture, protection and harvesting techniques in an integrated forest management strategy.

Valuable Hardwood Plantations

Long-rotation, slow-growing but valuable hardwood species have special technical properties, such as strength, natural durability, hardness and easy machining and appearance (grain, figure, texture, colour and other aesthetic qualities) that make them suitable for high-value end-uses such as furniture.

These high-grade hardwoods contrast with short-rotation, fast-growing, lesser-quality woods used for woodfuel, pulpwood or reconstituted products and less demanding building timbers. In tropical countries teak (*Tectona grandis*), mahogany (*Swietenia* spp.) and rosewood (*Dalbergia* spp.) are the main hardwood plantation species, while in temperate countries oak (*Quercus* spp.), ash (*Fraxinus* spp.), cherry (*Prunus* spp.), walnut (*Juglans* spp.), tulipwood (*Jacaranda* spp.) and hard maple (*Acer* spp.) predominate.

Because many valuable hardwood species are difficult to establish because of their ecological requirements or disease or insect susceptibility, focus has been on the easier species to grow, including teak (*Tectona grandis*), Indian rosewood (*Dalbergia sissoo*) and mahogany.

In 1995 the global areas of these species were 2 254 000, 626 000 and 151 000 ha, respectively. They accounted for about 10 per cent of total hardwood plantations in the tropics. More than 90 per cent of teak plantations were located in Asia, mainly in Indonesia, India, Thailand, Bangladesh, Myanmar and Sri Lanka. About 95 per cent of rosewood plantations are located in India and Pakistan. The largest mahogany plantations are located in Indonesia and Fiji, which together make up about 80 per cent of the established area.

The market preference for large piece sizes, slow growth and very long rotation lengths combine to reduce the attractiveness for commercial investment in these species. This is only partially counteracted by their value. The low return on capital investment, coupled with the long wait period for this return, has made it difficult to interest private investors without supportive, secure and stable government policies.

Table. Characteristics of Valuable Hardwoods used in Tropical Areas

Use categories	*Desirable wood properties*	*Main end-uses*	*Matching valuable hardwood species*	*Comments*
Decorative timbers	Appearance, consistent quality, dimensional stability, durability, good machining, staining and finishing properties	Quality furniture and interior joinery	*Tieghemella* spp.; *Entandophragma cylindricum*, *Chorophora* spp., *Aucoumea klaineana*, *Afrormosia* spp., *Entandophragma utile*, *Mansonia* spp., *Lovoa* spp., *Khaya* spp., *Swietenia* spp., *Dalbergia* spp., *Aningeria* spp.	Highest value, competitio n from temperate hardwoods and MDF
High to very high-density timbers	Appearance, strength, high natural durability, availability in large sizes	Principall y in constructi on	*Dipterocarpus* spp., *Lophira* spp., *Chlorophora* spp., *Ocotea rodiaei*	Small share of total tropical timber use
Low to medium-density utility timbers	Appearance, clear grain, natural durability, good machining properties	External joinery, shop fittings, medium-priced furniture	*Shorea* spp, *Hevea brasiliensis*, *Terminalia* spp., *Heritiera* spp.	Most commonly used, prone to competitio n from substitutes

As markets demand a continuity of supply, plantations need to be on a sustainable scale within a region. Some of the less common species are not known in the marketplace. Other potential market problems are that the timber may be wrongly associated with tropical deforestation and changing fashions that often occur with decorative timbers. Niche marketing is important for valuable hardwoods.

Projections for supplies of timber from existing valuable hardwood plantations indicate that because of the age class distribution and long rotations there will not be a significant increase in supply in the next 20 years. Future promotion of quality hardwood plantations needs to emphasize choice of species with versatile end-uses, market research and development to hold on to niche markets and maintained high standards from production to marketing. Careful site selection, use of high-quality planting materials of superior genetic origin and good silviculture are important. Planting programmes should be economically viable, environmentally appropriate and socially desirable. Incentives may also be necessary to stimulate private investment because of the long rotations.

Even though valuable hardwood plantations have the potential to reduce the pressure on natural forests, they will not prevent deforestation resulting from agricultural encroachment. The supply of large quantities of high-value timber could perhaps undermine the value of natural forest stands and so lead to more rapid destruction. Hence it is advisable, where possible, to manage plantations and forest resources and forest products in a complementary manner.

Plantations and Wood Energy

Woodfuels from plantations or natural or semi-natural forests are particularly important in developing countries, providing about 15 per cent of their total energy demand. Woodfuel provides about 7 per cent of energy demand for the world as a whole and in industrialized countries only 2 per cent. Woodfuel provides more than 70 per cent of energy needs in 34 developing countries and more than 90 per cent in 13 countries. Woodfuel makes up about 80 per cent of total wood use in developing countries and about 89 per cent in Africa

The prediction of a woodfuel crisis in developing countries in the 1980s was based largely on looking at supply and demand from forest plantations and natural forests. The reaction to the expected woodfuel crisis was to plant trees for this purpose, often in the form of traditional plantations. Many programme failures resulted from lack of appreciation for the complexities of bioenergy supply and demand, failure to take into account social aspects and people's needs and poor programme structures. The importance of planted trees on farmland, in villages and homesteads and along roads and waterways as a source of woodfuel supply was underestimated.

Rural communities harvest stems, branches, stumps, twigs, leaves and litter for woodfuels in chronic woodfuel supply areas. In these instances the nutrient recycling process is broken, resulting in degradation of forest plantation sites. In many rural communities in developing countries, woodfuel is considered a public free good, to be foraged from public natural and plantation forests. Often women and children collect the woodfuel at little or no cost. As a result, the growing of private forest plantations specifically for woodfuel, in which development costs and rotation cycles are involved, can be a foreign concept.

Asian studies show that forest-based supply can range from 13 per cent in the Philippines to as high as 73 per cent in Nepal. In many countries less than 50 per cent of fuelwood is from forests. Globally, non-industrial forest plantations in 1995 were estimated to cover about 20 million hectares. This was almost 17 per cent of the world's total plantation area in 1995. A significant proportion of these plantations were planted for woodfuel and 98 per cent were in developing countries. These plantation figures do not account for trees planted outside the forest on farms or in villages, etc., nor do they consider plantations that were considered agricultural plantations, such as *Hevea* or palm plantations.

In developing countries about one-third of the total plantation estate was grown primarily for woodfuel in 1995. Three-quarters of these plantations were in Asia, where they accounted for 60 per cent of total plantation production. In Latin America more than half of plantation production went to woodfuel; in Africa and Oceania a larger proportion of plantation production was as industrial wood. However, plantations, in general, provided only a small proportion of total woodfuel used. Uruguay is an interesting exception.

Production of woodfuel from plantations currently makes only a small contribution to energy requirements, although it is very important in some localities and countries. Plantations currently supply 5 per cent of woodfuel. Production from these non-industrial plantations is likely to double over the next 20 years, even with little expansion in area, because the age class distribution is heavily concentrated in young plantations. In an optimistic scenario where planting continues at the same rate as in the past ten years and then gradually declines, a 350 per cent increase in woodfuel production would be anticipated by 2020. By-products from wood-using industries will also contribute to increased fuelwood supply. The situation is less positive in Africa, where for a few countries declines are projected in plantation-based woodfuel production.

New Sources of Fibre

Since FRA 1990, advances in wood utilization technology have resulted in increasing importance of new sources of fibre - rubber (*Hevea brasiliensis*), coconut palm (*Cocos nucifera*) and African oil palm (*Elaeis guineensis*) - especially in the Southeast Asian subregion. These species account for 9.7, 12.0 and 6.0 million hectares of plantations, respectively. All grow in the humid tropics. In terms of plantation area, Asia has 92 per cent of the world's rubber, 86 per cent of the world's coconut palm and 78 per cent of the world's African oil palm. Indonesia,

Thailand and Malaysia have almost three-quarters of the rubber plantations; Indonesia and the Philippines have about half the coconut resources; and Malaysia has 55 per cent of the oil palm resource. All three species are grown principally for other products rather than wood, so when overmature they are available for fibre-based industries at minimal cost. Rubberwood is harvested when latex productivity declines and yields 100 m^3 per hectare of roundwood, but recovery for lumber is only 25 to 45 per cent because of poor form and small size. Most of the planted stands in Southeast Asia are owned by smallholders and are geographically dispersed, with poor accessibility and poor-quality stems. Currently the major proportion of industrially utilized rubberwood comes from large-scale plantations. Quality furniture, parquet, panelling, reconstituted panels, general utility timber and woodfuel, including charcoal, are made from rubberwood. However, the rubberwood must be processed within days of harvesting to minimize sapstain attack. The most developed downstream industries are in Malaysia, where the production of sawn rubberwood timber rose from 88 000 m^3 in 1990 to 137 000 m^3 in 1997 and medium density fibreboard (MDF) production from rubberwood reached 1.16 million cubic metres per annum by 1999. Exports of rubberwood furniture have grown from about US$74 million in 1991 to US$683 million in 1998. Rubberwood has become a substitute for light tropical forest hardwoods. Its acceptance as a sustainable plantation-grown, environmentally friendly timber has given it wide appeal.

Coconut palms are harvested as the copra yields decline (beyond 60 years) and yield 90 m^3 per hectare of coconut wood. Coconut palm has variable properties and is intrinsically difficult for conversion but can yield a relatively low-cost, general-utility timber for construction, panelling, stairs, door jambs, furniture, flooring and power poles. In 1993 Indonesia had 65 million cubic metres of overmature coconut stems which needed disposal before replanting. There is increasing interest in this raw material in European and North American markets. It is unlikely to replace conventional timber, but likely to find its way into niche markets. It will continue to be used as a low-cost construction timber.

Oil palm plantations are harvested for fibre beyond the 25 to 30 year rotations and yield about 235 m^3 per hectare. It is estimated that over 1.6 billion cubic metres of fibre will be available in the years to come from established resources in Southeast Asia. From 1996 to 1999 the area increased by 18 per cent. In Malaysia the area has increased by 3 million hectares in the past 30 years. Most oil palm plantations (unlike rubber and coconut) in the main growing countries, Malaysia and Indonesia, are managed by plantation companies or cooperatives. Oil palm by-products such as kernel shells, pressed fibres and empty fruit bunches are currently used in heat generation at the extraction plants. Water in the stems can reach five times the weight of dry matter.

The high moisture content as well as the high amounts of parenchyma tissue rich in sugar and starches make conversion into quality forest products a challenge. An MDF plant in Malaysia is currently being planned to utilize oil palm stems.

Plantation Substitutes for Natural Forest Products

With growing concerns about the status and loss of natural forests, the rapid expansion of protected areas and large areas of forest unavailable for wood supply, plantations are increasingly expected to provide substitutes for products from natural forests, particularly in Asia and the Pacific. In Asia and the Pacific it is estimated that 52 per cent of natural forests are not available for wood harvest because they are inaccessible or uneconomic to exploit. Of the unavailable forest in the region, it is estimated that about 38 per cent is legally reserved. In addition, logging bans have been imposed on large areas of natural forest covering about 10 million hectares. The reasons for these bans vary but were related to deforestation and forest degradation causing environmental problems in Thailand, the Philippines and China and to conservation requirements in Sri Lanka and New Zealand. As a result of the net effect of deforestation and removal of natural forests from wood production, some areas in the Asia and the Pacific region have wood deficits and roundwood harvesting is exceeding sustainable levels of cut. The worst affected areas are South Asia and insular Southeast Asia, with continental Southeast Asia also under strain. In contrast, New Zealand has surplus plantation wood available for export.

Of six examples studied in the Asia and the Pacific region, New Zealand is more than self-sufficient in wood production based on plantations. In China and Viet Nam, the importance of plantations will increase as planted resources mature. There have been serious problems with implementing plantation development programmes in Sri Lanka, the Philippines and Thailand. In Sri Lanka, India and elsewhere in the tropics, trees outside the forest are playing a critical role in roundwood and woodfuel supply.

Most countries in the region are becoming importers of wood, with imports expected to rise. Sometimes logging bans have shifted the problem to other countries. Problems with acquiring large areas of land in some countries make it difficult to implement industrial plantations. In the Philippines, Thailand and Viet Nam there have been social conflicts with local indigenous people or between traditional forest use and development, as well as between the rich and the poor. Sometimes incentives and the development of social forestry programmes are being used to help resolve such problems.

While it is clear that plantations will have an increasingly significant role in substituting products from natural forests, the impact will be felt on a case-by-case basis as governments and investors determine where and how plantations can be technically, economically and socially feasible as well as environmentally friendly. In the near term, plantations in Asia and the Pacific can make a contribution but cannot replace harvests from natural forests. It is likely that both in the region and globally the current pace of industrial plantation development will barely keep pace with losses from deforestation and transfer of natural forests to protected status. While it would be theoretically possible, actual plantation development is at present not sufficient to offset both growing consumption and declining harvest from natural forests.

Plantations and Carbon Sequestration

In the past ten years, the development of forest plantations as carbon offsets has evolved towards a market mechanism, although an organized market with carbon prices defined according to supply and demand forces is still a long way off. The adoption of the Kyoto Protocol in 1997 triggered a strong increase in investment in plantations as carbon sinks, although the legal and policy instruments and guidelines for management are still debated.

A number of countries have already prepared themselves for the additional funding for the establishment of human-made forests. The 1997 Costa Rica national programme was the first to establish tradeable securities of carbon sinks that could be used to offset emissions and the first to utilize independent certification insurance.

To date, greenhouse gas mitigation funding covers about 4 million hectares of forest plantations worldwide. The recognition of afforestation and reforestation as the only eligible land use, land use change and forestry activities under the Clean Development Mechanism of the Kyoto Protocol, as agreed in Bonn during the second part of the Sixth Conference of the Parties to UNFCCC in July 2001, will lead to a steep increase in forest plantation establishment in developing countries. The sink decision of the Bonn Agreement is expected to funnel additional funds into forest activities in developing countries and thus to strengthen the international efforts in this field.

However, it will also require a monitoring and verification system to ensure that these plantations will not be established at the expense of the local population or efforts to conserve biological diversity. Thus the decisions taken in Bonn to make the Kyoto Protocol ratifiable will also bear new challenges for forest plantation development.

CONTRIBUTIONS OF FOREST PLANTATIONS TO WOOD SUPPLY

The continuing increase in the area of forest plantations which have been established for industrial wood supply has been to meet the reduction of outturn foreseen from natural forests arising from deforestation and changes in land use (largely in the tropics and subtropics) or from natural forest being taken out of production and devoted to service functions such as conservation. It has been believed that the outputs from forest plantations can help to reduce the pressure on natural forests as sources of industrial wood supply; while the logging of tropical natural forests is not the prime cause of deforestation, logging roads often provide the means for farmers to gain access to forests. The reduction of logging, combined with effective protection, may thus help to reduce deforestation in certain locations until land use and ownership are clarified. However, none of these palliatives will remove the underlying causes of deforestation: high rates of population growth, poverty, hunger and a shortage of fertile land to cultivate.

The potential of forest plantations to meet demand for industrial roundwood is considerable; it has been estimated that the present global demand for paper pulp could be met from an area equivalent to only 1.5% of the world's closed forest area. No global estimates of current output of timber from forest plantations are available, although FAO's global fibre supply model (GFSM) estimated that the potential annual growth of industrial wood from forest plantations in developing countries was about 5% of the increment of natural forests in 1995.

In some countries, plantation production already makes a highly significant contribution to the industrial wood supply, for example in New Zealand 99% of industrial roundwood in 1997 was grown in plantations, while in Chile the equivalent figure was 95%, in Brazil and Argentina 60%, and in Zambia and Zimbabwe 50%. Estimating the future contribution of forest plantations to wood supply is at present imprecise and is based on many more or less unreliable assumptions, particularly concerning the rate at which afforestation will continue. By the year 2010 the GFSM (op. cit.) estimated that the potential increment from forest plantations would be about 40% of that from natural forests in Asia, Oceania and Latin America and about 15% in Africa, under rates of deforestation and afforestation largely the same as today.

FOREST VERTEBRATE RESOURCES: DIVERSITY AND DISTRIBUTION

Vegetational diversity and structural complexity have long been correlated with animal diversity (MacArthur & MacArthur 1961; Karr & Roth 1971; Urban & Smith 1989; Hansen & Hounihan 1996). Vertical habitat complexity and heterogeneity of regeneration phases in old-growth forests, compared with structurally simpler early successional forests, provide a variety of foraging, nesting and roosting sites and diverse microclimatic conditions that support high vertebrate diversity by niche segregation.

In northern temperate old-growth forests, elevated light levels in canopy gaps provide vertebrates with highly productive forage patches compared with younger forests that have continuous dense canopies. Large standing and fallen dead trees are far more common in old-growth than young regenerating forest and also provide a range of foraging and nesting opportunities. Foliage volume may be related to avian diversity by affecting the abundance of food resources or nest sites. However, correlation of forest structural complexity with bird species diversity has proved to be highly variable among regions and causal mechanisms linking forest structural complexity with bird diversity remain to be found. Mobility allows vertebrates to exploit temporally variable food resources and many, particularly mammals, are relatively generalist in their food requirements.

Consequently, vertebrates are able to respond behaviourally to changes in the distribution of ephemeral resources by localized habitat shifts, expansion of the home range or switching to alternative food sources. Behavioural responses to regional changes in resource availability can affect local abundance of migratory birds and mammals. High mobility buffers forest vertebrates (with some exceptions) from resource-associated breeding failure and extreme fluctuations of population size. However, extensive and prolonged anthropogenic or natural disturbances can affect vertebrate population dynamics by reducing clutch size or causing temporary suspension of breeding. Some resources have importance disproportionate to their abundance, often because they are available at times of food scarcity, and these may be particularly relevant to breeding success and population viability. Fig trees provide such 'keystone' resources to frugivorous birds and mammals in some tropical forests, and loss of fig trees is predicted to have wide-ranging impacts on the vertebrate communities of these forests.

Issues of Scale

Vertebrates are highly mobile and use resources on much larger spatial scales than plants or invertebrates. Foraging ranges of vertebrates may encompass several hundred hectares or more, and habitat requirements for nesting and feeding may be in quite different locations. Local disturbances may therefore be less important than large-scale disturbance phenomena in affecting vertebrate communities, and habitat patterns across landscapes should be more relevant to management strategies directed to forest vertebrates. Thus vertebrates are more likely to respond to habitat fragmentation and the heterogeneity and relative abundance of different habitat types than to forest structure or local complexity. Many vertebrates use resources from a variety of habitat types that occur in natural forested landscapes. These habitat types include a range of successional stages that are topographically, edaphically and hydrologically distinct vegetation zones. Human land uses tend to simplify or homogenize landscapes. Forest fragmentation leads to habitat isolation, reduced patch size and increased importance of edge effects as well as, more obviously, habitat loss. Single forest patches in anthropogenically fragmented landscapes are unlikely to contain the variety of habitat types and resources previously represented in the continuous forest landscape.

Vertebrates that use resources from several forest habitat types are therefore susceptible to changes in resource availability at the landscape scale. Single species operate at a variety of spatial scales during their life history. Thus many small birds defend nesting territories measured in square metres, use foraging ranges extending to several hectares, yet cover many kilometres during seasonal migrations. Even non-migratory species use resources, with varying strengths of interaction, at a range of spatial scales, and incorporating habitat features at multiple scales has been an important planning requirement for conservation of vertebrates such as the spotted owl *Strix occi-dentalis* in the forests of the Pacific North-West of the USA. In fragmented landscapes the extent of disruption depends on the scale of fragmentation relative to the mobility of the organisms and on the nature of the habitat matrix in which the fragments are located. Vertebrates that feed on unpredictable patchy and ephemeral resources, such as nectarivores and frugivores, operate on large spatial scales and tend to be more mobile than folivores or insectivores of similar size. Mobility facilitates movement between patches and consequently offers a degree of protection from the effects of habitat fragmentation, provided the matrix in which resource patches are located is favourable for movement and relatively risk-free. The nature of the habitat matrix and persistence of the resource is particularly important for vertebrates that use temporally and spatially predictable but ephemeral resources that require seasonal migrations between habitat zones. Seasonal habitat shifts are widespread among birds and mammals in response to climate, the availability of food resources, and nesting requirements. For example, bears in North America feed on berries in forest habitats but move to streams to feed on salmon during the annual spawning runs. Sparse berry crops can increase the movement of bears to streams, while a good fruit crop reduces the size of their home range. Forest patches surrounded by a hostile habitat matrix (or internal fragmentation features such as fences and roads) can impede vertebrate movement that, in addition to limiting access to resources, isolates subpopulations and prevents genetic mixing. By isolating subpopulations fragmentation increases local extinction probabilities by reducing the size of effective breeding population below thresholds of genetic and demographic viability. Provision of habitat corridors and careful planning of the land-use regime can increase landscape connectivity and facilitate greater movement of vertebrates between patches. Recent years have seen an increase in studies describing the responses of vertebrates to habitat fragmentation. These have been accompanied by theoretical developments, most notably describing thresholds of fragmentation beyond which the effects of isolation and small patch size increase rapidly.

The concepts of scale and heterogeneity in natural ecosystems are highly relevant to the integration of ecological principles with land management. The size and distribution of habitat patches need to be considered relative to the management objectives, be they conservation of particular species, continued sustainability of ecological processes, or both. Land managers will be aided by remote sensing techniques and geographic information systems (GIS) in the assessment of the suitability of landscapes to support high biodiversity, although these approaches need to be combined with studies of life history traits and species requirements. Variation in the frequency and size of natural disturbance events should be accounted for in management planning by incorporating or mimicking natural disturbance regimes at several spatial and temporal scales. Such planning needs to take a hierarchical perspective to match the mobility and requirements of biological entities that vary in size from small arthropods to herds of large mammals.

Bibliography

A.K. Mandal and G.L. Gibson: *Forest Genetics and Tree Breeding*, CBS Publication, Delhi, 2002.

A.R. Dabholkar: *General Plant Breeding*, Concept Publication, Delhi, 2006.

Bahar A. Siddiqui and Samiullah Khan: *Plant Breeding Advances and in vitro Culture*, CBS Publication, Delhi, 1997.

C.B. Singh: *Encyclopaedia of Plant Breeding*, Anmol Publication, Delhi, 2010.

D. Thangadurai, T. Pullaiah and Pedro A. Balatti: *Genetic Resources and Biotechnology, Vol. 2*, Regency Publication, Delhi, 2005.

Denis J. Murphy: *Plant Breeding and Biotechnology : Societal Context and the Future of Agriculture*, Cambridge University Press, New York, 2013.

Dharmendra Singh and S. Manivannan: *Genetic Resources of Horticultural Crops*, International Book Distributing Co, Delhi, 2009.

E.B. Babcock: *Genetics and Plant Breeding*, Agrobios Publication, Delhi, 2013.

Hari Har Ram and Rakesh Yadava: *Genetic Resources and Seed Enterprises : Management and Policies (2 Vols-Set)*, New India Publishing Agency, Delhi, 2007.

Kendall R. Lamkey and Michael Lee: *Plant Breeding: The Arnel R. Hallauer International Symposium*, Blackwell Publication, 2006.

L.D. Vijendra Das: *Problems Facing Plant Breeding*, CBS Publication, Delhi, 2000.

Lingaraj Patro: *Forest Genetic Resources Conservation and Management*, Mangalam Publication, Delhi, 2012.

Ludwig: *Dictionary of Plant Breeding*, Ivy Publication, Delhi, 2008.

M.P. Singh and Sunil Kumar: *Genetics and Plant Breeding, Vol. I and II*, APH Books, Delhi, 2009.

N.S. Kute and A.R. Aher: *Principles of Plant Breeding*, Agri-Biovet Press, Delhi, 2013.

P.V.G.K. Reddy: *Genetic Resources of Indian Major Carps*, Daya Publication, Delhi, 2005.

R.L. Kapoor and M.L. Saini: *Plant Breeding and Crop Improvement (2 Vol-Set)*, CBS Publication, Delhi, 1997.

Ranjeet K. Singh: *Genetic Resources and Biotechnology*, Surendra Publication, Delhi, 2010.

Robert W. Allard: *Principles of Plant Breeding*, Wiley India, 2010.

S.K. Gupta: *Practical Plant Breeding*, Agrobios Publication, Delhi, 2010.

S.K. Kataria: *Plant Breeding : Theory and Techniques*, Educational Publishers, Delhi, 2011.

Soumendra Chakraborty and Tapash Dasgupta: *Principles and Plant Breeding Methods of Field Crops in India*, New Delhi Publishers, Delhi, 2011.

T. Saravanan and K. Vanangamudi: *Principles and Methods of Plant Breeding*, International Book, Delhi, 2005.

Veenu Agarwal: *Genetics and Plant Breeding*, Oxford Book Company, Delhi, 2012.

Virendra Batra: *Genetics and Plant Breeding*, Oxford Book Company, Delhi, 2009.

Index